DEVELOPMENT AND MANAGEMENT OF REGULATORY OVERSIGHT FOR OPERATION OF A FIRST NUCLEAR POWER PLANT

The following States are Members of the International Atomic Energy Agency:

AFGHANISTAN	GAMBIA	NORWAY
ALBANIA	GEORGIA	OMAN
ALGERIA	GERMANY	PAKISTAN
ANGOLA	GHANA	PALAU
ANTIGUA AND BARBUDA	GREECE	PANAMA
ARGENTINA	GRENADA	PAPUA NEW GUINEA
ARMENIA	GUATEMALA	PARAGUAY
AUSTRALIA	GUINEA	PERU
AUSTRIA	GUYANA	PHILIPPINES
AZERBAIJAN	HAITI	POLAND
BAHAMAS	HOLY SEE	PORTUGAL
BAHRAIN	HONDURAS	QATAR
BANGLADESH	HUNGARY	REPUBLIC OF MOLDOVA
BARBADOS	ICELAND	ROMANIA
BELARUS	INDIA	RUSSIAN FEDERATION
BELGIUM	INDONESIA	RWANDA
BELIZE	IRAN, ISLAMIC REPUBLIC OF	SAINT KITTS AND NEVIS
BENIN	IRAQ	SAINT LUCIA
BOLIVIA, PLURINATIONAL	IRELAND	SAINT VINCENT AND
STATE OF	ISRAEL	THE GRENADINES
BOSNIA AND HERZEGOVINA	ITALY	SAMOA
BOTSWANA	JAMAICA	SAN MARINO
BRAZIL	JAPAN	SAUDI ARABIA
BRUNEI DARUSSALAM	JORDAN	SENEGAL
BULGARIA	KAZAKHSTAN	SERBIA
BURKINA FASO	KENYA	SEYCHELLES
BURUNDI	KOREA, REPUBLIC OF	SIERRA LEONE
CABO VERDE	KUWAIT	SINGAPORE
CAMBODIA	KYRGYZSTAN	SLOVAKIA
CAMEROON	LAO PEOPLE'S DEMOCRATIC	SLOVENIA
CANADA	REPUBLIC	SOUTH AFRICA
CENTRAL AFRICAN	LATVIA	SPAIN
REPUBLIC	LEBANON	SRI LANKA
CHAD	LESOTHO	SUDAN
CHILE	LIBERIA	SWEDEN
CHINA	LIBYA	SWITZERLAND
COLOMBIA	LIECHTENSTEIN	SYRIAN ARAB REPUBLIC
COMOROS	LITHUANIA	TAJIKISTAN
CONGO	LUXEMBOURG	THAILAND
COSTA RICA	MADAGASCAR	TOGO
CÔTE D'IVOIRE	MALAWI	TONGA
CROATIA	MALAYSIA	TRINIDAD AND TOBAGO
CUBA	MALI	TUNISIA
CYPRUS	MALTA	TÜRKİYE
CZECH REPUBLIC	MARSHALL ISLANDS	TURKMENISTAN
DEMOCRATIC REPUBLIC	MAURITANIA	UGANDA
OF THE CONGO	MAURITIUS	UKRAINE
DENMARK	MEXICO	UNITED ARAB EMIRATES
DJIBOUTI	MONACO	UNITED KINGDOM OF
DOMINICA	MONGOLIA	GREAT BRITAIN AND
DOMINICAN REPUBLIC	MONTENEGRO	NORTHERN IRELAND
ECUADOR	MOROCCO	UNITED REPUBLIC OF TANZANIA
EGYPT	MOZAMBIQUE	UNITED STATES OF AMERICA
EL SALVADOR	MYANMAR	URUGUAY
ERITREA	NAMIBIA	UZBEKISTAN
ESTONIA	NEPAL	VANUATU
ESWATINI	NETHERLANDS	VENEZUELA, BOLIVARIAN
ETHIOPIA	NEW ZEALAND	REPUBLIC OF
FIJI	NICARAGUA	VIET NAM
FINLAND	NIGER	YEMEN
FRANCE	NIGERIA	ZAMBIA
GABON	NORTH MACEDONIA	ZIMBABWE

The Agency's Statute was approved on 23 October 1956 by the Conference on the Statute of the IAEA held at United Nations Headquarters, New York; it entered into force on 29 July 1957. The Headquarters of the Agency are situated in Vienna. Its principal objective is "to accelerate and enlarge the contribution of atomic energy to peace, health and prosperity throughout the world".

IAEA-TECDOC-2033

DEVELOPMENT AND MANAGEMENT OF REGULATORY OVERSIGHT FOR OPERATION OF A FIRST NUCLEAR POWER PLANT

INTERNATIONAL ATOMIC ENERGY AGENCY
VIENNA, 2023

COPYRIGHT NOTICE

For further information on this publication, please contact:

Regulatory Activities Section
Nuclear Security of Materials and Facilities Section
International Atomic Energy Agency
Vienna International Centre
PO Box 100
1400 Vienna, Austria
Email: Official.Mail@iaea.org

© IAEA, 2023
Printed by the IAEA in Austria
December 2023

IAEA Library Cataloguing in Publication Data

Names: International Atomic Energy Agency.
Title: Development and management of regulatory oversight for operation of a first nuclear power plant / International Atomic Energy Agency.
Description: Vienna : International Atomic Energy Agency, 2023. | Series: IAEA TECDOC series, ISSN 1011-4289 ; no. 2033 | Includes bibliographical references.
Identifiers: IAEAL 23-01639 | ISBN 978-92-0-153423-1 (paperback : alk. paper) ISBN 978-92-0-153523-8 (pdf)
Subjects: LCSH: Nuclear power plants — Safety measures. | Nuclear power plants — — Management. | Nuclear power plants — Design and construction.

FOREWORD

The overall objective of regulatory oversight is to ensure that all safety related activities performed by the licensee throughout the lifetime of a nuclear installation demonstrate compliance with safety and security requirements and standards. Development and implementation of an effective regulatory oversight capability for the operation of a first nuclear power plant is one of the important considerations in ensuring the safety of the plant through the systematic application of an oversight mechanism in all phases of siting, design, commissioning and operation. To fulfil the objective of ensuring safety by incorporating all important aspects of regulatory oversight, a structured approach needs to be adopted.

The lessons identified from different IAEA expert missions, peer reviews, regulatory conferences and training workshops have highlighted that most of the countries embarking on a nuclear power programme are facing various challenges in the development and implementation of regulatory oversight in line with the phases of implementation of nuclear power programmes specified in IAEA Safety Standards Series No. SSG-16 (Rev. 1), Establishing the Safety Infrastructure for a Nuclear Power Programme.

This publication is intended to support Member States by providing practical information to develop and implement a structured regulatory oversight programme for the operation of a first nuclear power plant by identifying challenges and providing corresponding practical guidance from experienced Member States. The process is further elaborated by sharing good practices in the respective regulatory processes and actions taken to address those challenges.

This publication draws on the experience of experts from different regulatory bodies and provides a broader view of regulatory oversight of nuclear installations in Member States. The IAEA is grateful to all those involved in the development process. The IAEA officers responsible for this publication were Z.H. Shah and T. Hussain of the Division of Nuclear Installation Safety and K. Horvath of the Division of Nuclear Security.

EDITORIAL NOTE

CONTENTS

1. INTRODUCTION

1.1. BACKGROUND

The development and implementation of an effective regulatory oversight capability for the operation of nuclear installations is one of the important elements in authorizing initial operation and continued safe and secure operation of a nuclear power plant (NPP). The development of the necessary regulatory oversight capability requires a structured approach with due consideration of a number of elements including: implementation of a comprehensive regulatory framework; demonstrable leadership and management for safety and security together with the implementation of the management system within the regulatory body; development and maintenance of the competency of regulatory staff and any technical support organizations (TSOs); established interfaces with the licensee and other interested parties; sustained policy and financial support from the government, as well as support from the safety and security authorities in the vendor country; authorization and verification of trustworthiness for key plant operating personnel; and plans for managing the transition between the construction and operation phases, taking account of lessons learned.

The overall objective of regulatory oversight is to ensure that all safety related activities performed by the licensee throughout the entire lifetime of a nuclear installation comply with safety and security requirements and standards. It is important to ensure that a structured approach is followed for regulatory oversight for safety and security including all the important considerations documented in the IAEA Safety Standards Series and IAEA Nuclear Security Series publications and the use of international experience feedback, especially during authorization for operation of the NPP for the first time.

However, the results of some IAEA peer reviews and advisory missions to embarking countries revealed that there is a need for additional information to the Member States on development and implementation of regulatory oversight for the initial operation of the first NPP based on IAEA safety standards and IAEA nuclear security guidance.

Paragraph 4.26 of IAEA Safety Standards Series No. GSR Part 1 (Rev. 1), Governmental, Legal and Regulatory Framework for Safety [1], requires that:

> "The regulatory process shall be a formal process that is based on specified policies, principles and associated criteria, and that follows specified procedures as established in the management system. The process shall ensure the stability and consistency of regulatory control and shall prevent subjectivity in decision making by individual staff members of the regulatory body. The regulatory body shall be able to justify its decisions if they are challenged. In connection with its reviews and assessments and its inspections, the regulatory body shall inform applicants of the objectives, principles and associated criteria for safety on which its requirements, judgements and decisions are based."

Paragraph 4.28 of GSR Part 1 (Rev. 1) [1] states further that:

> "There shall be consistency in the decision making process of the regulatory body and in the regulatory requirements themselves, to build confidence among interested parties."

IAEA Nuclear Security Series No. 13, Nuclear Security Recommendations on Physical Protection of Nuclear Material and Nuclear Facilities [3], states in Fundamental Principle D:

"The State should establish or designate a competent authority which is responsible for the implementation of the legislative and regulatory framework, and is provided with adequate authority, competence and financial and human resources to fulfil its assigned responsibilities. The State should take steps to ensure an effective independence between the functions of the State's competent authority and those of any other body in charge of the promotion or utilization of nuclear energy."

Furthermore, IAEA Nuclear Security Series No. 13 [3], recommends in para. 3.18:

"The State's competent authority should have a clearly defined legal status and be independent from applicants/operators/shippers/carriers and have the legal authority to enable it to perform its responsibilities and functions effectively."

This TECDOC serves to provide useful information to regulatory bodies to help to ensure efficient and effective regulation for operation of the first NPP by taking into account international safety standards and nuclear security guidance, national framework for safety and security, and experience feedback of construction and commissioning oversight. The consideration of nuclear safeguards is an important aspect to consider during the management of regulatory oversight for the operation of the first NPP, however, safeguards aspects are beyond the scope of this TECDOC.

1.2. OBJECTIVE

The objective of this publication is to support the development and implementation of a structured regulatory oversight programme for the operation of a first NPP to ensure safety and security throughout its lifetime. The TECDOC provides relevant information that is useful to embarking countries by considering the challenges and offering suggested approaches for regulatory oversight of a first NPP.

1.3. SCOPE

The TECDOC is intended to be used by regulatory bodies for systematic development of regulatory oversight for construction and operation of the first NPP. The TECDOC addresses the important aspects of the regulatory oversight for construction and operation by considering the following:

- Organizational structure including clear interfaces with the licensee;
- Management system integrating all the regulatory processes;
- Assessment and provision of competencies and resources required for regulatory oversight and its sustainability;
- Approaches and practices to ensure effective control and follow up of regulatory oversight during operation;
- Authorization, review and assessment and inspections of compliance;
- Key aspects and steps to be considered to develop and strengthen regulatory oversight of human and organizational factors;
- Selection of important areas of regulatory oversight;
- Transition plan from oversight for construction and commissioning to oversight of operation;
- Communication with interested parties the regarding successful completion of construction and authorization for operational status of the installation.

1.4. STRUCTURE

This publication is divided into four sections and three annexes.

Section 1 provides information on the background, objective, scope and structure of the said publication. Section 2 summarizes the IAEA requirements, recommendations and guidance on managing regulatory oversight for the operation of NPPs. Section 3 provides information about essential elements of regulatory oversight through the early phases of NPP implementation. Section 4 details the challenges embarking countries may face during development and implementation of the mandate and core functions of the regulatory body and respective suggestions on approaches that may be adopted to address these challenges.

Annex I contains diverse case studies describing regulatory experience in Member States that are expanding their nuclear power programmes. Annex II lists the selected good practices from the IAEA's databases of Integrated Regulatory Review Service (IRRS) missions and International Physical Protection Advisory Service (IPPAS) missions that are relevant to the challenges and suggested approaches discussed in the TECDOC. Annex III describes lessons learned from experience with the COVID-19 pandemic as a reflection on coping with unforeseen situations.

2. RELEVANT IAEA REQUIREMENTS, RECOMMENDATIONS, GUIDANCE AND INFORMATION FROM OTHER PUBLICATIONS

This section summarizes the requirements, recommendations and guidance established in IAEA safety standards and other IAEA publications related to subject matter. IAEA Safety Standards Series No. SSG-16 (Rev. 1), Establishing the Safety Infrastructure for a Nuclear Power Programme [4], states in para. 2.1:

> "A nuclear power programme is a major national undertaking requiring careful planning and preparation, and a major investment in time and human and financial resources. While nuclear power is not unique in this respect, it is considered to be different because of the safety issues associated with the possession and handling of nuclear material and the long term commitment to ensuring safety after the decision to embark on a nuclear power programme has been made."

The phase of embarking on a first NPP may begin at different starting points ranging from no experience, to experience with laboratory scale nuclear facilities and industrial applications, operation of research reactors, or handling large amounts of radioactive material. In all of the aforementioned cases, the (human and financial) resources needed to secure a competent and fully functional regulatory body for the regulatory oversight of the first NPP are considerable. Therefore, the development of the regulatory body needs to be planned and implemented at an early stage of the programme. To conduct the regulatory oversight of the construction, commissioning, and early operation activities for the first NPP, the regulatory body needs to develop an extensive set of specialized competencies and processes to ensure an informed decision making process. INSAG Series No. 22, Nuclear Safety Infrastructure for a National Nuclear Power Programme Supported by the IAEA Fundamental Safety Principles [5], provides more information on the attributes of an independent regulatory decision making process.

In addition, para. 1.2 of SSG-16 (Rev. 1) [4] states:

"A considerable period of time is needed to acquire the necessary competences and to foster a strong safety culture before constructing and operating a nuclear power plant. While the prime responsibility for safety rests with the operating organization, the State has the responsibility to create a robust framework for safety upon committing itself to a nuclear power programme, which demands significant investment. Establishing a sustainable safety infrastructure is a long process, and it has been internationally acknowledged that a period of 10–15 years under optimum conditions is generally necessary between the consideration of nuclear power as part of the national energy strategy and the commencement of operation of the first nuclear power plant."

From a nuclear safety standpoint, the lifetime of a nuclear power plant is divided into five phases. Indicative average durations for each of these phases are as follows [4, 5]:

- Phase 1 is 'Safety infrastructure before deciding to launch a nuclear power programme' (average duration: 1–3 years);

- Phase 2 is 'Safety infrastructure preparatory work for construction of a nuclear power plant after a policy decision has been taken' (average duration: 3–7 years);

- Phase 3 is 'Safety infrastructure activities to construct a first nuclear power plant' (average duration: 7–10 years);

- Phase 4 is 'Safety infrastructure during the operation phase of a nuclear power plant' (average duration: 40–60 years);

- Phase 5 is 'Safety infrastructure during the decommissioning and waste management phases of a nuclear power plant' (average duration: 20 to more than 100 years).

These phases have been further elaborated in Fig. 1. SSG-16 (Rev. 1) [4] uses the same approach in considering Phases 1, 2, 3 and 4. Additional case studies, which further describe the experience of embarking countries, can be found in IAEA-TECDOC-1948, Experiences of Member States in Building a Regulatory Framework for the Oversight of New Nuclear Power Plants: Country Case Studies [6].

This TECDOC is consistent with the following IAEA publications:

IAEA Safety Standards Series No. SF-1, Fundamental Safety Principles [7], establishes the fundamental safety objective and ten associated safety principles, and briefly describes their intent and purpose. The fundamental safety objective — to protect people and the environment from harmful effects of ionizing radiation — applies to all circumstances that give rise to radiation risks. The safety principles are applicable, as relevant, throughout the entire lifetime of all facilities and activities, existing and new, utilized for peaceful purposes, and to protective actions to reduce existing radiation risks.

IAEA Safety Standards Series No. GSR Part 1 (Rev. 1), Governmental, Legal and Regulatory Framework for Safety [1], establishes requirements in relation to the governmental, legal and regulatory framework for safety. It covers the essential aspects of the framework for establishing a regulatory body and taking other actions necessary to ensure the effective regulatory control of facilities and activities utilized for peaceful purposes.

IAEA Safety Standards Series No. GSR Part 2, Leadership and Management for Safety [8], defines the requirements that support Principle 3 of the Fundamental Safety Principles in relation to establishing, sustaining, and continuously improving leadership and management for safety and an integrated management system. It emphasizes that leadership for safety, management for safety, an effective management system and a systemic approach (i.e. an approach in which interactions between technical, human and organizational factors are duly considered) are all essential to the specification and application of adequate safety measures and to the fostering of a strong safety culture.

IAEA Safety Standards Series No. GSG-12, Organization, Management and Staffing of a Regulatory Body for Safety [9], provides recommendations on meeting the requirements of GSR Part 1 (Rev. 1) [1], in relation to the organizational structure, management and staffing of the regulatory body. It addresses the arrangements and processes regulatory bodies need to consider in carrying out their responsibilities and functions efficiently and effectively and in an independent manner. It also provides guidance on an integrated management system to run the regulatory body.

IAEA Safety Standards Series No. GSG-13, Functions and Processes of the Regulatory Body for Safety [10], provides recommendations on meeting the requirements of GSR Part 1 (Rev. 1) [1], on the regulatory body's core functions and associated regulatory processes. It also assists authorized parties and others dealing with radiation sources in understanding regulatory procedures, processes and expectations.

IAEA Safety Standards Series No. SSG-16 (Rev. 1), Establishing the Safety Infrastructure for a Nuclear Power Programme [4], provides guidance on the establishment of a national nuclear safety infrastructure as a key component of the overall preparations required for emerging nuclear power programmes. It provides recommendations, presented in the form of 197 sequential actions, on meeting the applicable IAEA safety requirements during the first three phases of the development of a nuclear power programme. It is intended for use by persons or organizations participating in the preparation and implementation of a nuclear power programme, including government officials and legislative bodies, regulatory bodies, operating organizations and external support entities.

The Fukushima Daiichi Accident, IAEA Report by the Director General [11], provides a description of the accident and its causes, evolution and consequences, based on the evaluation of data and information from a large number of sources available at the time of writing. Reference [11] is of use to national authorities, international organizations, nuclear regulatory bodies, NPP operating organizations, designers of nuclear facilities and other experts in matters relating to nuclear power, as well as the wider public. It contains information on accident chronology, possible impact on the public and associated protective actions in national and international sphere.

IAEA Report on Strengthening Nuclear Regulatory Effectiveness in the Light of the Accident at the Fukushima Daiichi Nuclear Power Plant [12], provides an overview of the actions taken by nuclear regulators worldwide in the aftermath of the Fukushima Daiichi accident. It addresses actions taken by regulators to improve their own technical and organizational arrangements, actions requested by regulators from the licensees, general results and regulatory implications from these actions.

IAEA Nuclear Security Series No. 20, Objective and Essential Elements of a State's Nuclear Security Regime [13], provides nuclear security fundamentals, recommendations, and supporting guidance for Member States to assist them in implementing new nuclear security regimes, or in reviewing and, if necessary, strengthening existing ones. The publication is aimed at national policy makers, legislative bodies, competent authorities, institutions and individuals involved in the establishment, implementation, maintenance or sustainability of a State's nuclear security regime.

IAEA Nuclear Security Series No. 13, Nuclear Security Recommendations on Physical Protection of Nuclear Material and Nuclear Facilities (INFCIR/225/Revision 5) [14], provides guidance to States and their competent authorities on how to develop or enhance, implement and maintain a physical protection regime for nuclear material and nuclear facilities, through the establishment or improvement of their capabilities to implement legislative and regulatory programmes. The recommendations presented in this publication reflect a broad consensus among IAEA Member States on the requirements which should be met for the physical protection of nuclear materials and nuclear facilities.

IAEA Nuclear Security Series No. 19, Establishing the Nuclear Security Infrastructure for a Nuclear Power Programme [15], provides guidance on the actions to be taken by a State in implementing an effective nuclear security infrastructure for a nuclear power programme. The guidance provided is intended primarily for use by national policy makers, national legislators, competent authorities, institutions and individuals involved in the establishment, implementation, maintenance or sustainability of the nuclear security infrastructure for a nuclear power programme.

IAEA Nuclear Security Series No. 27-G, Physical Protection of Nuclear Material and Nuclear Facilities (Implementation of INFCIRC/225/Revision 5) [16], provides guidance and suggestions to assist States and their competent authorities in establishing, strengthening and sustaining their national physical protection regime and implementing the associated systems and measures, including operators' physical protection systems.

IAEA Nuclear Security Series No. 30-G, Sustaining a Nuclear Security Regime [17], addresses the sustainability of all aspects of a national nuclear security regime, including those relating to nuclear material and nuclear facilities, other radioactive material and associated facilities, and nuclear and other radioactive material out of regulatory control. The national level includes those elements of the nuclear security regime addressed by the State and its competent authorities that have general, State-wide applicability. The national level thus includes responsibility for: developing and implementing the overarching policy and strategy that support an integrated approach to nuclear security; developing and implementing the legislative and regulatory framework for nuclear security; assigning the roles and responsibilities for nuclear security; and defining the threat at the national level.

IAEA Nuclear Security Series No. 35-G, Security during the Lifetime of a Nuclear Facility [18], provides guidance to States, competent authorities, and operators on appropriate nuclear security measures during each stage in the lifetime of a nuclear facility, from initial planning of the facility through to its final decommissioning. This publication also addresses effective nuclear security in the transition between the stages.

IAEA Nuclear Security Series No. 14, Nuclear Security Recommendations on Radioactive Material and Associated Facilities [19], provides guidance to States and competent authorities on how to develop or enhance, implement and maintain a nuclear security regime for facilities that have with radioactive material and associated activities. This is to be achieved through the establishment or improvement of their capabilities to implement a legislative and regulatory framework to address the security of radioactive material, and of associated facilities and activities, in order to reduce the likelihood of malicious acts involving such material.

IAEA Nuclear Security Series No. 15, Nuclear Security Recommendations on Nuclear and other Radioactive Material Out of Regulatory Control [20], presents recommendations for the nuclear security of nuclear and other radioactive material that is out of regulatory control. It is based on national experience and practices and guidance publications in the field of security as well as the nuclear security related international instruments. The recommendations include guidance for States regarding the nuclear security of nuclear and other radioactive material that has been reported as being out of regulatory control as well as of material that is lost, missing, or stolen but has not been reported as such, or has been otherwise discovered. In addition, these recommendations adhere to the detection and assessment of alarms and alerts and to a graded response to criminal or unauthorized acts with nuclear security implications.

IAEA-TECDOC-1835, Technical and Scientific Support Organizations Providing Support to Regulatory Functions [21], describes the general characteristics, organizational aspects and types of services provided by TSOs to support regulatory functions and infrastructure in the Member States. As part of the organizational aspects, information is provided on the types of technical and scientific providers (e.g. internal or external to the regulatory body) and their respective challenges, and on the internal organization of TSOs to provide efficient and sustainable services and maintain expertise and competence. It covers all types of support for safety issues that may be provided by a TSO to a regulatory body to carry out its statutory functions, requiring a technical and scientific expertise in the nuclear and radiation safety field. Such support also applies to activities in related fields such as legal, training, and human resources.

IAEA Nuclear Security Series No. 23-G, Security of Nuclear Information [22], provides guidance on implementing the principle of confidentiality and on the broader aspects of information security (i.e. integrity and availability). It assists States in bridging the gap between existing government and industry standards on information security, the concepts and considerations that apply to nuclear security and the special provisions and conditions that exist when dealing with nuclear material and other radioactive material. Specifically, it seeks to assist States in the identification, classification, and assignment of appropriate security controls to information that could adversely impact nuclear security if compromised.

IAEA Nuclear Security Series No. 42-G, Computer Security for Nuclear Security [23], presents a detailed guidance on developing, implementing, and integrating computer security as a key component of nuclear security. This guidance applies to computer security aspects of nuclear security and its interfaces with nuclear safety and with other elements of a State's nuclear security regime, including the security of nuclear material and nuclear facilities, of radioactive material and associated facilities, and of nuclear and other radioactive material outside of regulatory control.

IAEA Nuclear Security Series No. 17-T (Rev. 1), Computer Security Techniques for Nuclear Facilities [24], serves as a guidance on how to establish or improve, develop, implement, maintain, and sustain computer security within nuclear facilities. It addresses the use of risk informed approaches to establish and enhance computer security policies, programmes; it describes the integration of computer security into the management system of a facility; establishes a systematic approach to identifying facility functions and appropriate computer security measures that protect sensitive digital assets and the facility from the consequence of cyber-attacks consistent with the threat assessment or design basis threat (DBT).

3. ESSENTIAL ELEMENTS OF REGULATORY OVERSIGHT FROM PHASE 2 TO PHASE 4

SSG-16 (Rev. 1) [4] sets out the main phases of a nuclear power programme and identifies some important safety steps for each phase. The first three phases are summarized in Fig. 1 [4].

INSAG-22 [5] identifies phase 2 as being critical for the establishment of the regulatory body. Once a national nuclear law has been adopted, that provides the regulatory body with a clear mandate and authority to carry out its mission, the regulatory body develops the work processes, human resources and competences needed to undertake its responsibilities in the nuclear power programme.

The development of the regulatory body is, thus, a high priority activity which requires 'strategic leadership' in phase 2 and continuing through phase 3.

The main tasks that are the responsibility of the regulatory body in phases 2 and 3 are complex technically and of a specialized nature. For example, the regulator has to be capable of:

- Issuing regulations and guides which set out the basis on which future licence applications will be assessed;
- Granting a construction licence following a thorough evaluation of the preliminary safety analysis report (PSAR) submitted by the licensee;
- Providing oversight of plant construction;
- Evaluating a final safety analysis report (FSAR) to support the issuance of an operating licence;
- Taking enforcement action against non-compliance.

The regulatory body also has to be ready to provide oversight of commissioning and operation of the NPP before the start of phase 4.

To perform these tasks, the regulator needs staff with competencies in a range of skills. Many of those areas are specific to nuclear power technology and safety and security, and so may not be immediately available within the embarking country. Therefore, the regulators in embarking countries have to start early in phase 2 to acquire the specialized areas of competence they will need to conduct their activities in phases 2, 3 and 4.

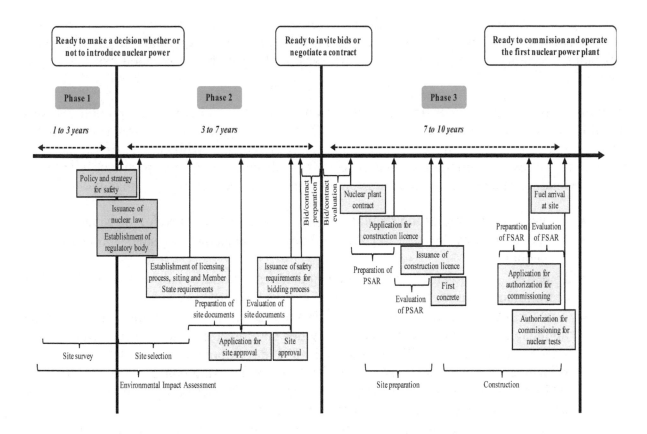

FIG. 1. The first three phases in a nuclear power programme [4].

The body of this TECDOC is structured around eight core themes, namely:

- Organization, staffing and competency;
- Core regulatory processes;
- Management system;
- Knowledge management;
- External TSOs;
- Emergency preparedness and response;
- Interfaces, communications and involvement of interested parties and the licensee;
- International cooperation.

These themes were identified through a systematic review and evaluation of relevant reference material, which was supplemented by the collective views and experience of the experts from Member States. These themes are the result of a consolidation of a range of potential concerns and issues which may be faced by the regulatory body during the various stages in the lifetime of the NPP. The main body of this TECDOC provides detail of the challenges and offers suggested approaches a regulatory body may adopt.

3.1. ORGANIZATION, STAFFING AND COMPETENCY

The staffing needs of the regulatory body are based on its mandate and core functions, as well as the general approach to regulation adopted by the Member State. The number of staff of the regulatory body and their specialized skills will also depend on decisions about the coverage of functional areas and the extent to which the regulatory body will use external experts and/or advisory committees. Irrespective of the arrangements in place, the regulatory body needs to have a sufficient number of staff with the knowledge, skills, experience and behavioural attributes necessary to operate the regulatory system along with any planned external expert support.

In view of the need for licensing and inspection knowledge and skills, the regulatory body should investigate opportunities for its staff to gain experience through cooperative arrangements with foreign regulatory bodies. Arrangements with those experienced in regulating the reactor technologies that the country will acquire would be particularly valuable [25]. Depending on the country's strategy, human resource needs in phase 2 may include:

- Strategic leadership of the regulatory body to establish and develop its capability;
- Expertise to develop the regulatory body's management system, processes and procedures;
- Expertise in regulatory organization and human resources development including recruitment, training, and contracting of external experts;
- Technical and regulatory expertise to develop and implement regulations and guides applicable to the proposed nuclear activities and to set out the basis on which future licence applications will be assessed. The development of regulations may be prioritized, for example those needed for the early licensing phases are developed first, while those required in following phases such as radioactive waste management and decommissioning are deferred for later development;
- Technical and regulatory expertise needed to review the site evaluation and environmental impact reports, if required as part of the licensing strategy;
- Expertise in stakeholder engagement and involvement.

Specific organizational capability requirements during phase 3 and early phase 4 include:

- Expertise needed to continue the evolution of the management arrangements; including the required safety analysis reports (SARs);
- Technical and regulatory expertise needed to review and assess the licence applications and capability to conduct effective oversight of the NPP construction;
- Recognizing the change in focus as the build of the NPP progresses, the resource capability needs, and the expertise for overseeing the commissioning and early plant operation.

Article 8 of the Convention on Nuclear Safety [26] obliges each Contracting Party to establish or designate a regulatory body entrusted with the implementation of the legislative and regulatory framework referred to in Article 7, and provided with adequate authority, competence and financial and human resources to fulfil its assigned responsibilities.

In addition, GSR Part 1 (Rev. 1) [1] puts the following requirements on the national government for establishment of a regulatory body:

"Requirement 3: Establishment of a regulatory body

The government, through the legal system, shall establish and maintain a regulatory body, and shall confer on it the legal authority and provide it with the competence and the resources necessary to fulfil its statutory obligation for the regulatory control of facilities and activities."

"Requirement 4: Independence of the regulatory body

The government shall ensure that the regulatory body is effectively independent in its safety related decision making and that it has functional separation from entities having responsibilities or interests that could unduly influence its decision making."

Paragraph 2.8 of GSR Part 1 (Rev. 1) [1] states:

"To be effectively independent from undue influences on its decision making, the regulatory body:

(a) Shall have sufficient authority and sufficient competent staff;

(b) Shall have access to sufficient financial resources for the proper and timely discharge of its assigned responsibilities;

(c) Shall be able to make independent regulatory judgements and regulatory decisions, at all stages in the lifetime of facilities and the duration of activities until release from regulatory control, under operational states and in accidents;

(d) Shall be free from any pressures associated with political circumstances or economic conditions, or pressures from government departments, authorized parties or other organizations;

(e) Shall be able to give independent advice and provide reports to government departments and governmental bodies on matters relating to the safety of facilities and activities. This includes access to the highest levels of government;

(f) Shall be able to liaise directly with regulatory bodies of other States and with international organizations to promote cooperation and the exchange of regulatory related information and experience."

INSAG Series No. 17, Independence in Regulatory Decision Making [27], notes that adequate and stable financing for all regulatory activities and their scientific and technical support is fundamental to independence in regulatory decision making and states that the financing mechanism should be clearly defined in the legal framework. If the costs of regulatory activities are to be reimbursed from the licensees, INSAG-17 [27] advises that the financing mechanism needs to be designed to prevent its misuse by licensees to reduce regulatory independence. Within its total budget, the regulatory body should have a high degree of independence in deciding how the budget is to be distributed between its various regulatory activities for the greatest effectiveness and efficiency.

The management of the regulatory body has the responsibility and authority to maintain sufficient staff with the necessary skills and expertise to carry out the regulatory functions discussed above. Such skills and expertise include regulatory competencies in four main areas identified in Safety Reports Series No. 79, Managing Regulatory Body Competence [28]:

- Legal, regulatory and organizational basis;
- Technical disciplines;
- Regulatory practices;
- Personal and behavioural.

The management of the regulatory body may employ one or more of the following subprocesses to acquire the needed competencies [28]:

- Analysis of competence needs:
  - Task analysis leading to determination of the necessary competences;
  - Analysis of existing competences within the regulatory body;
  - Gap analysis.
- Prioritization of competence needs and filling competence gaps:
  - Recruitment and human resources planning;
  - Staff training and development;
  - Management of external expert support;
  - Knowledge capture and management;
  - Reviews and audits of competence management and feedback.

During phases 2 and 3, the regulatory body will be proposing and promulgating safety and security regulations and guides to cover all foreseen nuclear activities, reviewing and assessing licensee applications, and developing and implementing the oversight programme for construction and operation activities. By the end of phase 3, the human resources for the regulatory body will need to be in place and competent to fulfil their functions for oversight of commissioning and operation in phase 4.

3.2. CORE REGULATORY PROCESSES

The organizational structure of the regulatory body may differ from Member State to Member State, depending on the national legal system and practices. The core functions to be performed by the regulatory body are:

(a) Development of regulations and guides (Regulatory framework);
(b) Review and assessment;
(c) Authorization;
(d) Inspection and enforcement.

3.2.1. Development of regulations and guides (Regulatory framework)

Requirement 32 of GSR Part 1 (Rev. 1) [1] states:

"The regulatory body shall establish or adopt regulations and guides to specify the principles, requirements and associated criteria for safety upon which its regulatory judgements, decisions and actions are based."

To meet this requirement, SSG-16 (Rev. 1) [4] recommends:

"The development of the regulatory framework involves maintaining a balance between prescriptive approaches and more flexible goal setting approaches. This balance might depend upon the State's legal system and regulatory approach. Since the approach chosen will have a major influence on the resources needed by the regulatory body, the persons expected to be in charge of the regulatory body should start learning and considering various regulatory approaches in Phase 1. A strategy should be developed to determine which regulatory approach will be chosen." (para. 2.71)

"During Phase 2, before the Member State decides which reactor technology is going to be deployed, the regulatory body should be aware of the two main alternative regulatory approaches: a prescriptive approach with a large number of regulations; or a goal setting approach that focuses on performance, functions and outcomes. Each regulatory approach has benefits and disadvantages, and there are also approaches that combine features of these two main alternatives. When a decision is made in Phase 3 on the reactor technology to be deployed, the regulatory body should adopt the approach that best suits the needs of the State. The regulatory body should have its chosen approach approved by the government, since there will be resource implications." (para. 2.80)

The regulatory body of an embarking country should consider establishing an organizational unit for the purpose of development of the regulatory framework required for the licensing of the NPP in different phases. The most knowledgeable people should be deployed to the task of developing regulations and guides, which form the basis of all the activities of the regulator. In developing regulations and guides, account should be taken of international standards and recommendations, obligations imposed by any conventions to which the State may be party, relevant industrial standards, vendor country regulations and any advances in technology. Consideration should also be given to regulations and guides from other States, as this may reduce the workload on the regulatory body in the drafting process.

In addition, if a country is only considering a single reactor design, it may be beneficial to base its regulations on those of the vendor's country. The advantages of this approach are that the embarking country's regulatory body might be able to finalize its regulations more quickly and the supplier would already be familiar with the regulations. The embarking country should also adapt the regulations to reflect specific national requirements and to ensure that the IAEA safety standards are adequately incorporated. If the regulatory body is not entirely self-sufficient, it may use external support for the development of regulations and guides, it may also adapt the international standards.

3.2.2. Review and assessment

Review and assessment are among the main continuous functions of a regulatory body. For a given review and assessment task, the services of a consultant or an internal or external TSO of the regulatory body may be used. Review and assessment often necessitate forming teams of specialists, depending on the complexity of the NPP under review and the scale and nature of the review and assessment work. Review and assessment should be carried out in accordance with principles and criteria set out clearly in specified review plans and procedures and based on the regulatory framework, codes and standards agreed by the regulatory body.

If the regulatory body is not entirely self-sufficient in all the technical or functional areas necessary to discharge its responsibilities for review and assessment, it should seek advice or assistance, as appropriate, from external experts (such as a dedicated TSO, universities or private consultants). Arrangements may be made to ensure that the consultants are effectively independent from the operator [29].

3.2.3. Authorization

In a nuclear power programme, the regulatory body verifies that site evaluation, design, construction, commissioning, operation and decommissioning comply with the relevant IAEA safety standards. The authorization process is the principal means by which the regulatory body is able to initially apply the legal and regulatory framework and by which the responsibilities of the applicant or authorized party are clearly connected to the legal framework. For complex facilities or activities and where the radiation risks are significant, the authorization process is usually referred to as a licensing process, which results in a licence in the form of a legal document issued by the regulatory body granting authorization to perform specified activities relating to the construction and operation of a facility or the conduct of an activity. The steps of the licensing process need to be discrete and follow a logical order as illustrated in Fig. 2 [29]. The regulatory body needs to keep records of authorization and retention of the relevant documents in connection with the authorization process.

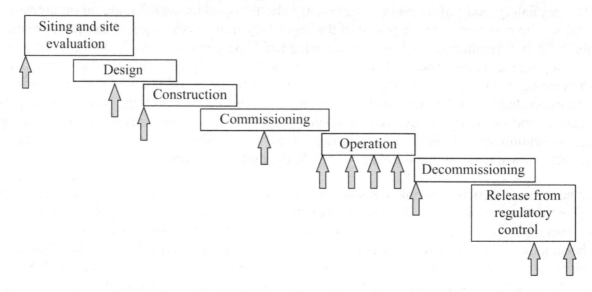

FIG. 2. Stages in the lifetime of a nuclear power plant [29].

3.2.4. Inspection and enforcement

Regulatory inspection is performed to make an independent check on the authorized party and the state of the facility or activity, and to verify that the authorized party is in compliance with the safety and security requirements prescribed or approved by the regulatory body.

A dedicated organizational unit for the coordination of inspection activities may be considered by the regulatory body. Inspections may concern particular aspects of an NPP and may be undertaken by individual inspectors or by teams of inspectors. The organization of inspections will depend on the scale of the activities and the availability of specialist personnel. If sufficient expertise is not available within the regulatory body, part of the inspection activities may be performed with the support of external experts. The regulatory body performs the inspections in accordance with its inspection programme, which consists of lower-tier detailed plans, administrative and technical procedures, and guidelines. These inspections are conducted during all stages in the lifetime of NPPs.

The principal objective of enforcement is to return the authorized party to full compliance with all relevant safety and security requirements and the authorization conditions, where non-compliances with safety and security requirements have been identified. This applies at all stages in the lifetime of an NPP (i.e. siting, design, construction, commissioning, operation, and decommissioning) or for the duration of an activity. Enforcement actions are intended to correct or improve any aspect of the procedures and practices of the authorized party or of the facility's systems, structures and components (SSCs), as necessary, to ensure safety and security. Enforcement actions may also include civil and criminal penalties and other sanctions which are designed to deter non-compliance.

3.3. INTEGRATED MANAGEMENT SYSTEM

An integrated management system is a set of interrelated or interacting elements (system) for establishing policies and objectives and enabling the objectives to be achieved in an efficient and effective manner [30].

Requirement 19 of GSR Part 1 (Rev. 1) [1] states:

"The regulatory body shall establish, implement, and assess and improve a management system that is aligned with its safety goals and contributes to their achievement."

The management system integrates safety, security and safeguards matters as well as other aspects such as the environment, society, and economy.

Furthermore, para. 4.15 of GSR Part 1 (Rev. 1) [1] states:

"The management system of the regulatory body has three purposes:

(a) To ensure that the responsibilities assigned to the regulatory body are properly discharged;

(b) To maintain and improve the performance of the regulatory body by means of the planning, control and supervision of its safety related activities;

(c) To foster and support a safety and security culture in the regulatory body through the development and reinforcement of leadership as well as good attitudes and behaviour in relation to safety on the part of individuals and teams."

The process based management system covers core regulatory processes and also management (executive) and support processes.

Requirement 8 of GSR Part 2 [8] states:

"The management system shall be documented."

In support of this requirement, para. 4.16 of GSR Part 2 [8] further states:

"The documentation of the management system shall include as a minimum: policy statements of the organization on values and behavioural expectations; the fundamental safety objective; a description of the organization and its structure; a description of the responsibilities and accountabilities; the levels of authority, including all interactions of those managing, performing and assessing work and including all processes; a description of how the management system complies with regulatory requirements that apply to the organization; and a description of the interactions with external organizations and with interested parties."

Paragraph 5.9 of GSG-12 [9] states:

"There are three different phases to an integrated management system:

(a) The development phase: Identifying and defining the processes necessary for the regulatory body to discharge its responsibilities and documenting in detail the content of each individual process in the context of the overall structure;

(b) The implementation phase: Implementing the processes in a planned and systematic way across the regulatory body;

(c) The maintenance phase: Ensuring that processes continue to be reliably applied and improved across the regulatory body."

However, the activities undertaken by the regulatory body will evolve during the different phases of a new nuclear power programme. Different activities, or processes, will have priority at certain times, such as development of regulations and guides, licensing of construction, and inspection of construction and commissioning. As a consequence, the regulatory body continues to update its management system. It may also decide to focus on the processes needed in the current phase and defer the development of those needed in later phases.

3.4. KNOWLEDGE MANAGEMENT

The regulatory body should acquire, manage, maintain, develop and preserve knowledge and information for building and maintaining adequate core competences. The objective should be to make informed decisions and to have competence to assess advice provided by advisory bodies, providers of external expert support, and information submitted by authorized parties and applicants as described in GSG-12 [9].

Knowledge management means an integrated, systematic approach to identifying, managing and sharing an organization's knowledge. Knowledge management helps the organization to better acquire, record, store and utilize knowledge [30].

Knowledge can be divided into 'explicit knowledge' and 'tacit knowledge'. Explicit knowledge is knowledge that is contained in, for example, documents, drawings, calculations, designs, databases, procedures and manuals. Tacit knowledge is knowledge that is held in a person's mind and has typically not been captured or transferred in any form. The knowledge management should cover both types of knowledge [30].

Knowledge is distinct from information: data yield information and knowledge are gained by acquiring, understanding and interpreting information. Knowledge and information each consist of true statements, but knowledge serves a purpose: knowledge confers a capacity for effective action [30].

Knowledge should be captured in a structured format that enables relevant information to be retrieved at the appropriate time to inform regulatory activities. Knowledge management system should be developed and maintained as a part of regulatory body`s management system. Knowledge management is closely linked to information management and document management. Paragraph 5.68 of GSG-12 [9] recommends:

"As part of its integrated management system, the regulatory body should establish a document management system that supports its information management processes, knowledge management processes and competence management processes."

Obtaining, creating and managing knowledge are challenging tasks in themselves. Paragraph 3.21 of GSG-12 [9] recommends:

"Processes should be established, from the early stages of development of the regulatory body's integrated management system, to acquire, use, maintain, store and retrieve information and knowledge. These processes should be supported by specific tools and techniques, for example:

- Questionnaires, interviews and discussions, and reports (special attention should be paid to the transfer of knowledge when experienced staff leave or retire from the regulatory body);
- Databases, libraries, 'knowledge portals' and archives."

3.5. EXTERNAL TECHNICAL SUPPORT ORGANIZATIONS

The regulatory body should have full competence to perform all regulatory functions during construction, commissioning and operation of an NPP. It may, however, be necessary for the regulatory body to use the services of external experts or an external technical support organization. External TSOs may be within that State or from a foreign State. In general, most of the countries embarking on a nuclear programme receive technical and scientific support from several TSOs, via bilateral agreements, projects, and contracts from the vendor's country or countries which have experience in regulating similar type of NPPs, from the IAEA, and from different regulators forums. In case of using outside entities or external experts, the regulatory body's personnel should have sufficient technical knowledge to enable them to identify problems, to determine whether it would be appropriate to seek assistance from an external expert, and to understand, evaluate and use any relevant advice from the external expert. Therefore, the regulatory body should have a process and procedures in place to obtain suitable external expert support to gain input that can be used in making regulatory decisions [4, 9, 27]. Further details on selection of external expert support are given in Appendix I of GSG-12 [9].

3.6. EMERGENCY PREPAREDNESS AND RESPONSE

Requirements for preparedness and response for a nuclear or radiological emergency are specified in IAEA Safety Standards Series No. GSR Part 7, Preparedness and Response for a Nuclear or Radiological Emergency [31], IAEA Safety Standards Series No. GSR Part 3, Radiation Protection and Safety of Radiation Sources: International Basic Safety Standards [32], and GSR Part 1 (Rev. 1) [1].

The government makes provision for emergency preparedness to enable a timely and effective response in a nuclear or radiological emergency, including the development of an appropriate national legal framework and the designation of responsibilities of response organizations.

The regulatory body makes each authorized party responsible for preparing an emergency plan and for making arrangements for emergency preparedness and response (EPR). Emergency arrangements include a clear assignment of responsibility for immediate notification of an emergency to the response organizations. The regulatory body takes account of the fact that, in an emergency, routine regulatory administration such as the issue of prior authorizations may need to be suspended in favour of a timely emergency response.

In preparing an emergency plan and in the event of an emergency, the regulatory body advises the government and response organizations and provide expert services (e.g. services for radiation monitoring and risk assessment for actual and expected future radiation risks) in accordance with the responsibilities assigned to it [1].

The State has to prepare a national response plan for nuclear security and develop the necessary systems and measures to respond to criminal or unauthorized acts with nuclear security implications involving nuclear or other radioactive material out of regulatory control [15].

3.7. INTERFACES, COMMUNICATIONS AND INVOLVEMENT OF INTERESTED PARTIES AND LICENSEE(S)

Within countries embarking on a new nuclear power programme, other governmental authorities may have responsibilities related to the regulation of new NPPs. These responsibilities may include areas such as:

- Safety of workers and the public;
- Protection of the environment;
- Emergency preparedness and response;
- Nuclear security;
- Safety in relation to water use and the consumption of food;
- Land use, planning and construction;
- Controls on the import and export of nuclear material, radioactive material and dual use items and technology.

Requirement 7 of GSR Part 1 (Rev. 1) [1] states:

"Where several authorities have responsibilities for safety within the regulatory framework for safety, the government shall make provision for the effective coordination of their regulatory functions, to avoid any omissions or undue duplication and to avoid conflicting requirements being placed on authorized parties."

This coordination and liaison can be achieved by means of memoranda of understanding, appropriate communications, and regular meetings. Such coordination assists in achieving consistency and in enabling authorities to benefit from each other's experience. In countries with no prior experience of nuclear power, it may fall to the new nuclear regulatory body to inform other agencies on specific requirements related to the NPP.

Establishing a framework for effective communication among all relevant stakeholder organizations (e.g. those responsible for nuclear safety, nuclear security, safeguards and facility operations) is a prerequisite for successful completion of the NPP project.

In addition, effective communication and involvement of the public and other interested parties is recognized as fundamental across the international nuclear community in order to contribute to building confidence and trust in the regulatory process.

Principle 2 of SF-1 [7] states in para. 3.10 that, among other aspects:

"The regulatory body must: ... — Set up appropriate means of informing parties in the vicinity, the public and other interested parties, and the information media about the safety aspects (including health and environmental aspects) of facilities and activities and about regulatory processes; — Consult parties in the vicinity, the public and other interested parties, as appropriate, in an open and inclusive process."

Further, Requirement 36 of GSR Part 1 (Rev. 1) [1] states:

"The regulatory body shall promote the establishment of appropriate means of informing and consulting interested parties and the public about the possible radiation risks associated with facilities and activities, and about the processes and decisions of the regulatory body."

The regulatory body should place requirements on authorized parties to inform and, when appropriate, consult interested parties about the radiation risks associated with the operation of a facility or the conduct of activities as described in IAEA Safety Standards Series No. GSG-6, Communication and Consultation with Interested Parties by the Regulatory Body [33]. Additionally, the regulatory body should coordinate with other competent authorities the review of the DBT at regular intervals and evaluate the consequence of any changes in the DBT.

A range of possible stakeholders in a country with a new nuclear power programme is listed in GSG-6 [33]:

- The general public including people living or working in the vicinity of the NPP;
- Licensees and regulated organizations;
- Governmental ministries and departments;
- Educational and research institutes;
- TSOs and professional organizations;
- Non-governmental organizations and special interest groups;
- Neighbouring countries.

3.8. INTERNATIONAL COOPERATION

International cooperation is essential for countries with nuclear power programmes, or for countries planning them. The organizations and persons involved in the utilization of nuclear energy are interdependent in that the performance of one may have implications for all, and a serious nuclear accident would be of major significance around the world. Recognition of this dependence has led to a number of international arrangements that are intended to enhance safety in all States [1]. The international cooperation brings some obligations to the state, but also provides invaluable help and support for promoting safety and security nationally and enhances international confidence and trust. For further details, SSG-16 (Rev. 1) [4] may be consulted.

Further in this section, different forms of international cooperation are discussed.

3.8.1. Global safety and security regime

A nuclear power programme in any State cannot be treated in isolation, owing to the potential transboundary effects of a radioactive release. States have a shared need for universal safe and secure operation of nuclear facilities and safe and secure conduct of activities. The national policies and the strategies adopted by the government therefore take full benefit of effective participation in the global nuclear safety and security regimes. This includes for example becoming party to international conventions and codes of conduct for safety and security and nuclear liability, establishing mechanisms for exchanging construction, operating and regulatory experiences with other organizations and creating mechanisms to inform neighboring countries about the planned nuclear power programme.

3.8.2. Cooperation with the IAEA and other organizations

The IAEA offers its Member States a wide array of services. In those review services, an IAEA-led multinational team of experts compares actual practices with IAEA safety standards. There are review missions both for regulators and for utilities, and information about these missions is available on IAEA website. Member States submit a formal request to the IAEA for a mission to be initiated. In addition, the IAEA offers a wide spectrum of education and training activities. These include face-to-face training courses and workshops, as well as online learning, fellowship programmes and schools on various nuclear related topics.

Other international organizations exist, too, for example the OECD Nuclear Energy Agency or regional associations such as in Europe WENRA (Western European Nuclear Regulators Association) or ENSRA (European Nuclear Security Regulators Association). Participating in the activities of these organizations can support the competence building of the newly established regulatory body.

3.8.3. Cooperation with the regulatory bodies of other countries

Multilateral and bilateral cooperation is one of the ways to achieve harmonized, high quality, and effective regulatory oversight. It is beneficial to establish cooperation with the regulatory body of the vendor country, but also with the regulatory bodies of countries licensing the same reactor type.

For new regulatory bodies, it is necessary to consider international cooperation to overcome the challenges which they may face at the early stage. These challenges, along with suggested approaches, are discussed in Section 4.8.

4. CHALLENGES AND SUGGESTED APPROACHES

This section captures the opinions of some Member States on some potential challenges based around the topics discussed earlier. These challenges may be faced by an embarking country as it develops and implements the mandate and core functions of the regulatory body. This section also offers some pragmatic suggestions on approaches that may be adopted to address these challenges. It is structured into subsections based on the same themes as addressed in Section 3; firstly, capturing generic challenges and suggested approaches, then identifying any specific challenges in each phase. There are also some 'statements of advice' included at the end of the section to assist Member States.

4.1. ORGANIZATION, STAFFING AND COMPETENCY

The regulatory body needs to ensure that it has adequate resources (both financial and human) and independence to enable the regulatory body to discharge its accountability under the legal and regulatory framework of the State. This section captures the key challenges in this area for each phase and suggested approaches to manage those challenges.

4.1.1. Generic challenges and suggested approaches

Challenge No. 1: Strategic approach to defining and developing the organizational capability of the regulatory body

A strategic approach is needed towards the development of the organizational capability of the regulatory body, for the current and future phases of the nuclear power programme. An appropriate strategy will enable the regulatory body to have the right people with the right skills, at the right time, supported by the right infrastructure and will also assist with the management of interested parties' expectations.

A significant challenge for the senior management of the regulatory body in a country that is embarking on a new nuclear power programme arises from the need to deliver products and services relevant for the current phase of the programme (for example regulations and licences), while developing the infrastructure needed for the next phase (for example inspection programmes) in line with the NPP project schedule.

Suggested approaches

- It is suggested that the regulatory body takes a strategic perspective to identifying key points of change throughout the entire lifetime of the NPP and makes plans to effectively manage these transitions. Such key points include: first, the move from review and assessment of the first construction licence application to inspection of construction and commissioning activities, and secondly, the transition from inspection of construction and commissioning to regulating the early operations of a NPP.
- The strategy needs to reflect the legal and regulatory framework of the Member State and the extent of the regulatory body's jurisdiction. This strategy may also define the core capability needed within the regulatory body and a methodology of how this capability will be established. Factors to be considered include:

- The extent of the nuclear programme and regulatory framework;
- Organizational vision, purpose and values;
- Organizational design principles and structure;
- Defining key roles and responsibilities, and the associated essential skills;
- Level of delegated authority within the regulatory body;
- The reactor technology(s) proposed to be deployed in the State;
- Core competencies requirements needed within the regulatory body and how these change during the different phases;
- The current relevant skills and knowledge residing within the State;
- Access to existing skills and knowledge, within the State or beyond, through TSOs (see Sections 4.4 and 4.5 for further considerations);
- Cooperation with the regulatory body of the vendor's country for competence development;
- Internal and third party review and evaluation (self-assessments / IAEA / other regulatory body / professional bodies).

- Project management tools are very helpful to produce a comprehensive strategic plan covering all aspects including human resource planning, staff recruitment and development, financial considerations, establishing infrastructure (for example, buildings, IT systems, and document control), interested parties and engagement, and it has to be used to manage the development of the organizational capability. It is advisable to produce a transition plan for each phase (for example, construction, supply chain oversight to commissioning oversight) and to effectively manage the transition. Ideally, this transition is managed as a project, using project management disciplines, to minimize the impact on front line regulation, while delivering the enhanced capability.
- The strategic plan may incorporate lessons learned reviews throughout the different phases, so improvements can be made to the management system.

Challenge No. 2: Developing and maintaining the competencies of the regulatory body for the relevant phases

Establishing and maintaining a competent regulatory body is a fundamental expectation across the international nuclear community. The challenge is to build the appropriate competency for the relevant phase, while recognizing and managing the changes in competencies during the period of transition between the phases.

Suggested approaches

- Establish and implement a competency framework to assess staff against the requirements for the relevant phase and identify the competency gaps.
- Develop a systematic training programme for the regulatory body, covering legal, technical disciplines, regulatory practice, and behavioural competencies to effectively implement the regulatory framework. This training programme includes periodic reassessment of regulatory staff to ensure the currency of their skills.
- To maintain a sustainable and competent regulatory body, introduce effective work force and succession planning processes. Ideally, these plans would cover a period of 5 years, informed by forward work force demands, potential staff turnover (including retirements) and a vulnerability analysis of the organizational capability, both core

staff and TSO. This exercise is normally conducted on an annual basis, to inform the management of the regulatory body.

- The availability of longer term work force planning projections enable informed resource (recruitment) plans to be produced and implemented, incorporating the range of skills and knowledge required. The resource plan considers a variety of sources for recruitment, including experienced staff, graduates and temporary staff with specialist skills required for a limited time, if needed.
- Staff retention has proved to be an issue in a number of Member States, recognizing the limited availability of skilled nuclear practitioners. The regulatory body normally gives due consideration to a retention policy for key roles, and produce and maintain succession plans to enable the organization to grow and improve resilience.
- Targeted use of TSOs, specialized contractors or organizations that have the relevant skills to fulfil the needs of the regulatory body can be helpful. If TSOs are used, it is recommended that knowledge transfer requirements are specified in the contracts, to enable the core capability of the regulatory body to mature.
- The regulatory body may choose to employ the TSOs and external contractors during specific periods in the programme, such as the review of major licence applications, thus enabling it to focus on building the core competencies needed in-house for oversight of NPP operation.

Challenge No. 3: Funding and independence of the regulatory body

Adequate and stable financing for all regulatory activities and their scientific and technical support is fundamental to maintaining independence in regulatory decision making.

Suggested approaches

- The financing mechanism for the regulatory body needs to be defined clearly in the legal framework.
- If the costs of regulatory activities are to be reimbursed from the licensees, then the financing mechanism needs to be designed to prevent its misuse by licensees to reduce regulatory independence.

Within its total budget, the regulatory body needs to have a high degree of independence in deciding how the budget is to be distributed between its various regulatory activities for the greatest effectiveness and efficiency.

Challenge No. 4: Leadership and organizational effectiveness

Effective leadership is an essential attribute of any effective organization. The leaders need to set the strategic direction for the organization by inspiring staff through a sense of common purpose, forward vision and organizational values. This leadership framework is then managed and delivered effectively through the arrangements within the organization.

Suggested approaches

- Effective leadership is important for the organization's performance and a regulatory body is no exception. The regulatory body has to consider developing a leadership model that exemplifies the principles and values of the organization. The regulatory body has to consider establishing a leadership development programme as part of increasing the effectiveness of the regulatory body and developing the potential of their staff.

- Conducting planned effectiveness reviews of the organizations capability is advisable. These reviews could be part of a self-audit programme and are likely to provide essential information to the leadership of the regulatory body.
- It is recommended that a regulatory body establish an integrated approach for learning and continuous improvement within the organization, which enables the organization to embrace learning and increase effectiveness, both from internal and external experiences.
- Enhanced leadership is essential through times of change, particularly when the management arrangements are still maturing through the various phases.

4.1.2. Phase 2 specific challenges

Challenge No. 1: Implementation of the strategic plan for the regulatory body

In phase 2, the regulatory body is established and therefore a significant challenge is the implementation of the strategic plan for the regulatory body and development of organizational capabilities.

Suggested approaches

- The regulatory body needs to consider developing an initial overarching strategy for establishing their organizational capability. This strategy will require engagement of interested parties but should be owned by the management of the regulatory body.
- The strategy ought to have sufficient granularity for the current phase or subphase and an indicative view of the longer term needs.
- The strategy needs to be periodically reviewed and refined as the organization matures.

Challenge No. 2: Setting security expectations

An essential element in this early phase is the establishment of arrangements for security vetting of personnel, both within the regulatory body and industry. These arrangements need to be developed so they appropriately reflect the activities being conducted in the phase.

Suggested approaches

- Security vetting needs to be established among personnel, from both regulatory body and industry, by the ministry responsible for national security.
- Special considerations need to be established for individuals from different States.
- The regulatory body needs to define and effectively communicate security expectations to all relevant parties.
- The regulatory body needs to develop an inspection programme to oversee the vendors' and licensees' arrangements for nuclear security, including personnel security.

Challenge No. 3: Protection of sensitive and proprietary information

Protection of sensitive information related to nuclear technology is crucial during the whole lifetime of an NPP because at some stage the regulatory body may consider sharing proprietary information to the vendor, the licensee (including contractors/suppliers), and the competent authorities. Therefore, protection of sensitive information related to nuclear technology is crucial during the whole lifetime of an NPP.

Suggested approaches

- The regulatory body and each competent authority ensure that starting from a very early phase of the project relevant personnel are trained in procedures for protection of such information and sensitive assets. Requirements have to be established for protecting the confidentiality of information.
- These requirements address limiting access to sensitive information to those whose trustworthiness has been established through a trustworthiness assessment (e.g. identity verification, criminal history review, polygraph testing etc.), appropriate to the sensitivity of the information, and those who have a need to know to perform their duties.
- The regulatory body develops an inspection programme to inspect licensee organizations and premises where sensitive information may be held to verify that they have the appropriate arrangements in place.
- The regulatory body also imposes requirements for protection of sensitive information upon its own TSOs and contractors with whom such information is to be shared.
- The regulatory body establishes agreements and systems to share information, with all the relevant parties, starting from a very early phase of the project, recognizing that each relevant group may have different requirements for the sharing of information.

4.1.3. Phase 3 specific challenges

Challenge No. 1: Transitioning skills specifically needed for phase 3

As the regulatory activities transition into phase 3, there is a significant change in the skills and experience needed to effectively regulate the supply chain, manufacture, construction, and commissioning of the NPP project.

Suggested approaches

- The regulatory body needs to proactively manage this transition by evolving the competency framework in advance of the need and reassessing staff against the requirements for regulatory oversight of construction and commissioning (including supply chain manufacturers).
- Staff training and development plans may be implemented prior to the need. This training may include knowledge transfer from the vendor on the design basis and any unique safety and security features. The regulatory body may consider cross-qualification of existing staff to ease the burden of hiring additional personnel.
- The targeted use of TSOs may be an option for the regulatory body to enhance its capability, although knowledge retention needs to be considered. This point is discussed later in this publication.

4.1.4. Early phase 4 specific challenges

Challenge No. 1: Cultural change within the regulatory body between phase 3 and early phase 4

It is important for the regulatory body to proactively develop and maintain a healthy safety and security culture within its organization which can be further enhanced during the transition period from phase 3 into phase 4. Experience has revealed that this is an area that is not to be underestimated and requires a change in attitudes and working methods for the start of operation. The development of organizational culture takes time.

Suggested approaches

- Strong leadership, management system, readiness review process and culture for safety and security are necessary elements to for smooth cultural transation between phase 3 and early phase 4.
- Consider approaches to foster effective teamwork within the regulatory body, for example regular 'keep in touch' meetings, briefings and engagement sessions and periodic workshops.
- The regulatory body, through its interactions with the licensee, could seek to positively influence the licensee's safety and security culture.
- Consider treating every action at the latest stages of phase 3 as if the NPP was in operation. This will provide ample time for the staff to adapt to the culture shift without having to worry about the consequences of an accident involving an operating nuclear reactor.
- In order to develop an appropriate safety and security culture programme for the regulatory body, benchmarking with other regulatory bodies is recommended.

Challenge No. 2: Managing the transition to regulating steady state operations

This is a significant period of change, which requires a mindset change by the organization, moving from regulating a dynamic project environment, to steady state operations.

Suggested approaches

- The leadership team of the regulatory body needs to provide effective support for its staff during this period, which is likely to include adoption of different engagement approaches, processes and reporting mechanisms, requiring the staff to employ new skills and behaviours.
- As the project transitions from phase 3 to early phase 4, the competency framework is reviewed and updated to ensure it reflects the requirements for overseeing steady state operational activities.
- It is also important to note that the licensee will be going through a similar change, so enhanced regulatory oversight may be required during the period.
- To emphasize this point, experience from other new build projects shows that pre-operations, commissioning, and preparation for fuel loading can be hectic and the time schedules can change constantly. The regulatory body has to have flexibility in its resource planning and allocation to be able to react proportionately to these inevitable changes.

- It is recommended that a targeted training and development plan may be developed and implemented to assist the regulatory body's staff through the transition, that focus on where the management system and core processes have been changed for the operational phase and the behavioural expectations associated with early and steady state permanent operations.
- It is important to create a strong nuclear safety and security culture for the regulatory body at an early stage — one culture for safety and security. In organizations where safety and security are critical, this one organizational culture would consider safety and security matters in all decision making processes.
- It is a good practice to benchmark a regulatory body's capability and performance against other comparable organizations, which could include other Member States' regulatory bodies. This would provide a useful indicator of the maturity of the organization.
- A regulatory body normally considers the benefit of establishing cooperation agreements with relevant research organizations who could provide useful third party technical insight into some of the challenges associated with the technology and emerging research.

4.2. CORE REGULATORY PROCESSES

Core regulatory functions are subject to graded approach consistent with the magnitude of the possible radiation risks arising from the facility or activity. During phase 2 to early phase 4, regulatory bodies may face many challenges, including but not limited to, difficulties with regard to the development and adoption of regulations and guides, shortage of trained staff, unavailability of standards in working language, difficulties in implementation of review schedule, coordination between different organizations, etc. This section highlights such challenges along with suggested approaches to cope with these challenges.

4.2.1. Generic challenges and suggested approaches

Challenge No. 1: Developed regulatory framework for the current phase

In an embarking country, the development pace of the operator can be aggressive, and the regulatory body has to keep itself level with the operator's pace. During this time, the regulatory body may have insufficient time to develop regulations, guides, programmes, policies and procedures.

Suggested approaches

- The regulatory body may consider adoption of technical standards of the vendor's country as part of its own regulatory framework; or use the regulatory framework of other Member States operating the same type of NPPs through bilateral arrangements. Another option is to adopt IAEA safety standards as the basis for the regulatory framework.

Challenge No. 2: Regulatory approaches

In case the regulatory body is unable to develop the regulations, guides and oversight programme, the services of a consultant from vendor's country may be hired, in such case the probability exists that different approaches and/or practices may be reflected in regulations and guides and the oversight programme.

Suggested approaches

- It is suggested to hire the consultant from the vendor's country for the development of regulations and guides or adopt the IAEA safety standards for licensing the first NPP. Hiring consultants from the vendor's country has the advantage that any regulations, guides and oversight programme developed will be based on a regulatory approach already practiced in the vendor's country.

Challenge No. 3: Shortage of trained and experienced human resources

The foremost challenge for the regulatory body of an embarking country is likely to be the lack of trained staff and the financial resources for performing core regulatory functions such as development of regulations, safety assessment, and inspection and enforcement. Trained and experienced resources are essentially required in all phases to perform regulatory functions in an efficient and effective manner.

Suggested approaches

- The regulatory body needs to prepare long term human resource development plans for recruitment of staff and their capacity building and update it well in advance of each phase. The regulatory bodies in some Member States submit this plan to the government for allocation of sufficient funds in each financial year. The workload on the regulatory body will increase during the transition from phase 2 to phase 3 and early phase 4 and may stabilize or decline in phase 4. Therefore, the regulatory body will prepare transition plans for shifting from one phase to another so that efficient and effective regulatory oversight activities can continue and be enhanced.

Challenge No. 4: Management of proprietary and confidential information

During the review and assessment process, when the reviewer requests the designer's documents necessary for the safety assessment, the designer may be reluctant to share the proprietary design documents with the regulatory body of the licensee's country.

Suggested approaches

- The regulatory body needs to highlight this issue with the licensee before signing of the contract so that the licensee can negotiate with the vendor the provision of proprietary design documents in an agreed working language. The regulatory body may establish appropriate regulations and agreements with its personnel and contractors for the protection of proprietary and confidential information from unauthorized use or disclosure.

Challenge No. 5: Regulatory oversight during unforeseen and unexpected circumstances (COVID-19 pandemic)

Situations of national emergencies may arise from different natural or man-made disasters without any warning. As the task of a regulatory body is continuous, a robust mechanism needs to be in place to deal with regulatory requirements during such emergencies. One such example is the COVID-19 pandemic that began in 2020. Lockdowns, travel restrictions and strict quarantine measures were imposed worldwide to prevent the spread of COVID-19. These public health and safety measures and the threat of the disease to the regulatory staff presented a steep challenge.

During a pandemic situation, a major challenge may be faced by the regulatory body as inspections may not be able to be conducted directly at the site. Moreover, no face-to-face meetings of the regulatory body with the licensee and its contractors/subcontractors are possible.

Suggested approaches

Under such pandemic circumstances, the following approaches may be used by the regulatory body for maintaining effective regulatory oversight during construction, commissioning, and initial operation:

- Selection of construction and installation control points may be limited, and the licensee or contractor may be asked to provide the record of activities along with photographic and video evidence of the activities. Additionally, live video technology can be utilized to perform remote inspections. Inspection activities may have to be prioritized based on risk significance.
- In order to minimize exposure to illness and to reduce the number of infections among regulatory staff, a policy of staffing reduction and rotation may be employed. Staff working from home may be given home based assignments such as review and assessment, preparation of procedures, etc.
- Inspectors may be tested for disease and then quarantined at a plant site for a few months to continue inspection during the installation and commissioning stage. Adequate resources may be provided to carry out these tasks effectively. The number of inspectors in quarantine may depend on the workload and/or commissioning stage.
- In order to reduce the load from quarantined inspectors during the commissioning stage, the licensee may submit the entrance meeting packages electronically. The entrance meeting packages may be reviewed by a commissioning test inspection team off the site, and queries sent to the licensee. After resolution of the queries, the updated documents along with checklists may be sent to quarantined inspectors at the site who could then conduct the required inspection activities.
- Video conference arrangements may be made with the licensee and its contractors/subcontractors for regular meetings to resolve technical issues. The use of modern technology and mobile apps may be adopted for better interaction and communication.
- After completion of the commissioning tests, the commissioning test reports may be submitted by the licensee electronically. The review of commissioning test reports may be carried out by commissioning test inspection team outside quarantine in coordination with site inspectors telephonically. After resolution of all queries with the licensee electronically, the acceptance of commissioning tests may be provided to the licensee.

4.2.2. Phase 2 specific challenges

Challenge No. 1: Use of national standards by the vendor

Although the major safety and security related codes and standards are agreed in the contract, during the design and construction phases the vendor may introduce some national industrial standards. Translated versions of these standards in an agreed working language may not be available for use by the regulatory body, if required during inspections at the construction site and manufacturing facilities.

Suggested approaches

- The regulatory body needs to highlight this issue with the licensee in phase 2 before signing of the contract so that the licensee can negotiate with the vendor for the provision of translation in an agreed working language. Regarding the use of national standards for design and construction of NPPs, the vendor may be requested to provide the equivalent of these standards with the international standards agreed with the licensee during the negotiation of the contract.
- The standards to be used during the construction and installation activities are agreed between the regulatory body and the operating organization at the time of agreeing to the codes for design (review and assessment). Agreements may also be needed in phase 2 to provide the national standards of the vendor country in an agreed language, or provision of their equivalent international standard. Where difference of opinion exists, the operator needs to demonstrate to the regulatory body that the same level of safety is being achieved with the alternative approach.

Challenge No. 02: Conducting vulnerability assessment

In phase 2, conditions which can influence the vulnerability of the new NPP should be considered during the siting process in addition to the nuclear safety considerations, e.g. topography and existing nearby facilities.

Suggested approaches

- During phase 2, the regulatory body evaluates any local or regional threats that could impact the NPP and topography of the candidate sites that may influence the vulnerability of the NPP (e.g. the possibility of a malicious act). Moreover, the effect of existing nearby nuclear or industrial facilities are also to be evaluated from the security point of view.

4.2.3. Phase 3 specific challenges

Challenge No. 1: Understanding of industrial standards of vendor's country

The regulatory body performing review and assessment of a SAR for the first time may not have a thorough understanding of the industrial standards referred to in the NPP design documentation. The comprehensive review and assessment of the licensee's submissions during the construction phase require industrial standards to be used as criteria to verify the results of design analyses and tests during manufacturing. Some of the standards are only available in the language of the vendor's country.

Suggested approaches

- The staff members assigned to the review could study the relevant standards and apply these standards during their first review process, if necessary, with the support of external experts, so that they are able to fully understand and apply these standards in the next phase of the review. The regulatory body may train its staff in the language of the vendor's country for better understanding of their national standards.

Challenge No. 2: Approval of modifications by the regulatory body

The licensee is required to get approval of plant safety related modifications from the regulatory body during design, construction, and manufacturing of equipment, as well as during installation and commissioning. However, to meet the project timelines at various stages, the contractors may prefer to perform modifications without the approval of the regulatory body.

During the commissioning phase of an NPP, experience has shown the majority of the issues reported are related to inconsistencies between actual system configurations on site and the information provided in the SAR and in system design manuals.

Suggested approaches

- The regulatory body establishes a process and criteria for a graded approach for control of design modifications, which may involve prior approval of safety significant changes and apply different regulatory oversight for less significant modifications (for example inspection, review of post-modification records and reports). The regulatory body performs field inspections to verify the installation of systems in accordance with the SAR and system design manuals. Any discrepancy observed is communicated to the operating organization seeking justification. Similarly, during the commissioning phase, several changes may be made in the design logics, operating setpoints, etc. to obtain the desired results. The operating organization is required to incorporate all these modifications in the SARs and other relevant documents. The updated SAR will be approved by the regulatory body before granting the operating licence.

Challenge No. 3: Licensing of main control room operators

In some countries, the regulatory body is responsible for granting licences to main control room (MCR) operators. In embarking countries, the staff of the regulatory body may not yet have the expertise and knowledge to conduct oral and practical (simulator) examinations.

Suggested approaches

- The regulatory body will identify well in advance a licensing examination committee preferably from those staff who are involved in the design review, commissioning programme review and performed inspections during construction, installation, and commissioning. The members of the licensing committee study in depth the assigned areas and familiarize themselves with the plant design to allow them to conduct oral examinations of the first batch of operators. The regulatory body may also seek opportunities to train some personnel on a similar plant in the vendor country along with operating personnel. Personnel involved in commissioning activities can be licensed and be involved in the operation of the plant, so that their knowledge and experience will be increased. Subsequently, these personnel can be involved in oral and simulator examinations of MCR operators.

- Alternatively, the regulatory body could adopt an approach of placing the responsibility on the licensee for training and qualification of operating staff. In this case, the regulatory body oversees the licensee's training and qualification process (for example delegate a staff member to the licensee's examination committee).

Challenge No. 4: Non-availability of end of manufacturing reports

Manufacturing reports of equipment are normally made available at the site along with the equipment. It is possible that some of the equipment will arrive at the site for installation without complete end of manufacturing reports. It will be a challenging task for inspectors to verify that the equipment has passed all factory-based manufacturing tests for comparison with commissioning test results at the site.

Suggested approaches

- The regulatory body establishes the requirements for a comprehensive quality assurance management programme which the licensee is required to implement for its own activities and for controlling the activities of its contractors and subcontractors.
- The regulatory body oversees the licensee's implementation of the quality assurance management programme, including availability of end of manufacturing reports for the equipment arriving at the site. The inspectors of the regulatory body may perform inspections on arrival of important safety class equipment and report any non-compliance to the licensee for its attention to focus on the rest of the equipment to be delivered.

Challenge No. 5: Review schedule of licensing submission

- The review schedules of the SARs required for the construction and operation licences are generally agreed between the regulatory body and the licensee. In some Member States, the review periods for granting a licence for construction, operation, etc. are defined in the national regulatory framework and hence timeline for completion of SARs review cannot be violated. However, implementation of the planned schedule is sometimes difficult due to the following reasons:

  - Incomplete submission of information in some chapters of the SAR by the licensee/designer;
  - Late delivery of responses from the licensee to the queries raised by the regulatory body at different stages of the safety review;
  - Unavailability of some equipment/systems qualification reports for review as the vendor considers them proprietary.

Suggested approaches

- Before agreement on the SAR review schedule, the expectations of a regulatory body are communicated with the licensee in order to consider the following:

    - Finalize the timeframe with the vendor (designer) for preparation of the SARs and submission to the licensee;
    - Consider the time required to check the completeness of SARs as per the requirements of the regulatory body before submission;
    - Finalize the review schedule of SARs in consultation with the vendor;
    - The regulatory body informs the licensee that (i) the review schedule depends on the quality, completeness, and timeliness of its submissions, and (ii) the regulatory body will endeavour to meet the agreed schedule but in case of conflict it will give priority to its responsibilities for assurance of safety and security.

Challenge No. 6: Configuration management during project phase

It is often very challenging for a licensee to cope with the requirements of the configuration management during the construction and installation phase because of the enormous amount of data of the NPP project, especially the documentation regarding modifications in the design of SSCs. This issue could arise due to lack of staffing and inadequate interfaces between the main contractor, the licensee/operating organization, and the designer.

Suggested approach

- The regulatory body may wish to verify that the licensee and its contractors have an effective configuration management process in place. The process ensures that all aspects of configuration management related to SSCs are covered by a detailed procedure and that any subsequent changes to the preliminary design go through assessment, approval, implementation and verification phases and all of these changes are traceable through the documentation. It is important for the licensee to be aware of all the important changes and be part of the modification process.

Challenge No. 7: Availability of simulator instructors

For the operation of the first NPP in the country, it would be a challenging task for the licensee to simultaneously develop a team of operators as well as the instructors for the simulator. It will also be a challenge for the regulatory body to confirm that the simulator instructors are well trained and available to provide the required level of training to the operators.

Suggested approaches

- The regulatory body may proactively engage with the licensee to gain assurance that plans are in place in terms of human and financial resources to train the operating staff and simulator instructors well ahead of time either in the vendor's country or in another country having a similar type of NPPs.

Challenge No. 8: Utilization of experience feedback of similar plant

At a site where more than one NPP is to be constructed, experience feedback of one unit may not be fully utilized by the contractor and the operator in the next unit due to involvement of different staff.

Suggested approaches

- The regulatory body gains confidence if the licensee has put in place effective processes to collect relevant learning and experience for the next unit of the NPP. The process should enable a categorization of learning and a mechanism for learning to be shared in future construction. This process captures learning both from the site and other relevant sources from across the nuclear sector. The regulatory body can conduct an administrative inspection of the licensee on this subject to confirm that an experience feedback mechanism is in place.

Challenge No. 9: Handling special security issues during phase 3

During the construction phase, a large number of persons enter the site which may pose additional threats related to unauthorized access, prohibited items or tampering with devices.

Suggested approaches

- During the construction phase, the regulatory body has to ensure that control measures are established for detection and prevention of unauthorized access to the construction site of vehicles, persons and packages. Moreover, during phase 3, the security measures are in place for preventing the introduction of contraband to the construction site. Additionally, the security measures prevent any tampering with facilities or equipment that could aid in the execution of a malicious act after the NPP becomes operational (for example keeping items in secure storage until installation to reduce the possibility for tampering).

Challenge No. 10: Potential threat originating from nearby operating nuclear power plant

Nearby operating facilities can cause additional threat because of the large number of employees and tools which can be used by an adversary.

Suggested approaches

- The regulatory body has to ensure that measures are in place for isolating construction activities from other nearby operational facilities (e.g. those located on the same site) that may cause additional vulnerabilities during phase 3.

Challenge No. 11: Drafting of construction licence and licensing conditions

At the time of issuance of the construction licence, some open issues may exist which require the commitment by the licensee to be completed during the construction phase. The regulatory body may issue the construction licence along with some associated licence conditions or additional hold points if there are no major safety related concerns.

Suggested approaches

- It is recommended that the organizational unit responsible for issuance of the licence tracks focuses on safety significant issues while drafting the licensing conditions. The licence conditions may be divided into generic and time specific licence conditions. After issuance of the licence, the regulatory body establishes a plan for compliance with the licence conditions in coordination with the unit of the regulatory body responsible for inspection activities.

4.2.4. Early phase 4 specific challenges

Challenge No. 1: Coordination between organizations during commissioning phase

During commissioning, different activities are happening in parallel and several organizations can be involved in the activities. Some plant systems may be under testing while others are already in operation, still under construction, or are in maintenance. Simultaneously, the development and finalization of the respective documentation (e.g. SAR update) is being undertaken. It is crucial to coordinate all the different activities. Typically, a work permit system is established for controlling the different works.

Different organizational units may have responsibilities for the different activities. Regardless of how the activities are organized, the roles and responsibilities have to be clear and sufficient communication has to be guaranteed. The MCR personnel are required to always be aware of the status of the systems and components and of planned and ongoing activities. Equipment damage or even incidents may occur due to lack coordination and inadequate communication.

Suggested approaches

- The regulatory body is required to verify that a proper process is in place for coordination of different activities during commissioning. During the routine inspection of MCR and commissioning tests, resident inspectors may verify the interface and coordination between different organizational units. Any deficiency noted is required to be communicated to the licensee as soon as possible.

Challenge No. 2: Non-availability of maintenance records

During commissioning tests, if on-the-spot maintenance work is undertaken by the vendor's personnel on certain equipment without following the work control process (thus no record of faults and corrective maintenance), the same problem may also occur again to said equipment during operation phase. In such cases, the operating organization may not find any record of corrective maintenance performed on the subject equipment previously by the vendor.

Suggested approaches

- The regulatory body normally ensures that the licensee has appropriate arrangements in place for the work control process to perform any maintenance activity on the equipment and the operating organization is required to ensure that all the maintenance records are transferred to it by the commissioning subcontractor at the time of plant handover.

4.3. MANAGEMENT SYSTEM

The management system of regulatory body is a tool to monitor the performance and effectiveness of regulatory business. It introduces a culture of continuous improvement.

The management system is developed and applied using a graded approach that reflects the significance and complexity of the conducted activity and the hazards, and the magnitude of the potential impacts (risks) connected with it.

4.3.1. Generic challenges and suggested approaches

Challenge No. 1: Understanding and acceptance of management system

It is important that the management and the staff of the regulatory body possess adequate knowledge and understanding of the nature of a management system, its purpose, and contribution to the effective performance of the organization. Experience has shown that, when this understanding is weak, the management system did not have the needed participation and motivation of the regulatory body's staff.

Suggested approaches

- The commitment and support of the regulatory body senior management is crucial.
- All regulatory body staff receive training by experienced personnel (possibly consultants) covering both theoretical background and practical explanation of the management system relevant to a regulatory body.
- Motivation to develop, implement and maintain the management system continuity is important.
- Involvement of relevant regulatory staff working in different units and utilization of external expertise (if required) while developing the management system is needed.
- Periodical review of the management system is a good practise. Involvement of senior management of the regulatory body in this process is essential.

Challenge No. 2: Demonstration of leadership for safety and security by managers of regulatory body

Visible and active support, strong leadership and the commitment to safety and security of the regulatory body management are fundamental to developing and maintaining strong safety and security culture. Each organization has its own organizational culture; some organizations can be managed informally while other ones are managed very formally. The way of demonstrating leadership for safety and security usually reflects the overall organizational culture. Weak or insincere management commitment to safety and security results in weak safety and security culture of the overall organization.

Suggested approaches

- The senior management of the regulatory body communicates through their own behaviour and management practices that safety and security are overriding priorities.
- A common understanding of safety and security culture, including awareness of radiation risks and hazards, is supported.

- An organizational culture that supports and encourages trust, collaboration, consultation and communication is supported.
- Reporting of problems and proposing countermeasures and improvements are encouraged.
- The safety–security interface is appropriately managed. Actions are taken on a continuing basis to foster a strong safety and security culture. Formal and informal mechanisms can be developed and used in accordance with the overall organization culture.
- Effective programmes dedicated to build a safety and security culture can be introduced in the regulatory body.
- Periodical management review of the management system includes assessment of leadership for safety and security and safety and security culture.
- Benchmarking and experience exchanges with similar organizations is helpful.
- Development of a team of experts by the regulatory body in the area of safety and security culture self assessment is important.
- Communication of results of the self-assessments of leadership for safety and security and of safety and security culture at all levels within the regulatory body contribute towards fostering and sustaining a strong safety and security culture.

Challenge No. 3: Efficient planning

Efficient planning of the development of the management system covering activities, responsibilities, capacities needed, timing etc., helps to achieve goals, distribute the effort evenly, justify in-time recruitment and training, etc. Not every organization is strong in planning generally and this can bring long term consequences in work organization and goals achievement. Moreover, once NPP construction starts and the project progresses to more advanced stages, the demands increase on the regulatory body staff. Planning can help to allocate the time and effort needed to develop and maintain the management system in-time.

Suggested approaches

- Planning of the management system development has to be coordinated with other plans and activities of the regulatory body. Because planning itself is a self contained discipline, it can be led by a person or team responsible for planning with the appropriate training and planning tools, like suitable software.
- It is recommended to plan the development of the management system in several stages which reflect the regulatory activities in different phases of the NPP programme. The management system for particular activities of the regulatory body has to be in place before those activities are required to be implemented.
- Reporting mechanisms are created to provide feedback on how the plan is fulfilled, how goals are achieved and what is the status of each planned activity. In case of substantial delays proper countermeasures are adopted.

4.3.2. Phase 2 specific challenges

Challenge No. 1: Good timing of management system development and resources provision

Experience has shown newly formed organizations can more readily implement a management system than those which have worked for a time without one. This happens because the staff in a new organization are open to new methods and have no loyalty to old ways of working. It can happen that the regulatory body is established and managed for several years without a documented management system because of lack of knowledge, human and financial resources, etc. During this period, certain management practices can become entrenched. Later on, it is necessary to integrate all those practices in a newly established management system and it can bring additional work and changes necessary to be in compliance with management system requirements.

Suggested approaches

- It is recommended to start with development of the management system during the establishment of the regulatory body (i.e. phase 1 or phase 2). This could be understood as a trial stage and as an opportunity to learn. Later, according to the management system development plan, the scope can be broadened and the management system can be developed step by step to the needed scope (phase 2 and phase 3). External support can be used to bring necessary knowledge and experience. Once the management system is implemented, it can be modified and improved based on measurement, assessment and improvement activities. Development of the management system can be managed as a project with defined activities, responsibilities, due dates, reporting mechanisms, etc. using project management tools.
- During the development of the regulatory body, it is recommended that benchmarking of other international regulatory bodies is adopted.

Challenge No. 2: Preparation and application of process based procedures

It is important to define the overall process model for the organization and the architecture of individual processes. It can happen that methodology is not prepared, or it is not sufficient (too simple or in the opposite very complicated). As a consequence, the management system is not developed in a unified way and, therefore, it is not efficient from the very beginning.

Suggested approaches

- It is better to identify the needed processes and to prepare a procedure for development of management system documents before elaborating other processes. If needed, an external consultant can be used to provide experience and guidance. Then, after elaboration of the first procedure for the selected processes, an evaluation of the method could be done (applicability, efficiency, etc.) and work on other procedures can continue.

4.3.3. Phase 3 specific challenges

Challenge No. 1: Coordination for first core regulatory process

It is quite common that the first robust review activity of the regulatory body is related to the construction licence application in phase 3. There is typically a significant number of regulatory body staff and possibly TSO staff that participate in the review and, therefore, a coordination role is crucial. A management system procedure for review covers many topics such as inputs, outputs, elaboration of review methodology and assessment criteria, workflows, communication with licensee, information management, etc. Supporting inspections can also be executed. A consistent and unified approach of all regulatory body staff during particular activities is needed.

Suggested approaches

- It is a good practice to establish a strong coordination team, which manages the review process, provides mentoring to all participating regulatory body personnel, executes internal surveillance of proper application of management system procedures and other documents, and supports communication with the licensee and other authorities involved in the licensing process. An external consultant with proper experience can be used to support such a team.

Challenge No. 2: Continuous changes to be reflected in the management system

Extensive changes in the regulatory body can happen before the first NPP is put into operation, with many modifications to processes, practices, and day-to-day activities to reflect the changing scope of regulatory body oversight. The capability to assure oversight during operation require new qualifications, new documentation, new approaches, new information management, new types of reporting and many other matters. Preparation lasts several months if not years. Due to the quantity of changes, and their potential temporary character in early phase 4, coupled with a lack of regulatory body staff capacity, frequent updates to the management system cannot be undertaken.

The tasks related to construction and commissioning and operation have different characteristics, and sometimes it happens that priority is given to immediate issues connected with construction and commissioning, while the preparation of matters for future needs is postponed and therefore delayed.

Suggested approaches

- It is a good practice to update the management system at reasonable intervals which reflect important stages (milestones) in regulatory body activities and regulatory body staff workload. In the meantime, some flexible managerial mechanisms with temporary validity can be used and documented.
- It is a good practice for the regulatory body to dedicate capacity toward preparation for operation oversight, including development of management system procedures and staff training, in order to assure that the regulatory body is ready to conduct oversight for early phase 4 when the authorization is granted.
- All activities relevant to readiness for oversight of operating reactors can be managed as a project or as a transition plan. The project or transition plan covers activities, schedules, resources, and responsible persons, and managed by appointed person or project manager. Status reporting and comprehensive reviews by the regulatory body management are essential. In case of substantial delays, proper countermeasures are adopted.

4.3.4. Phase 4 specific challenges

Challenge No. 1: Measurement, assessment, and improvement of the management system of regulatory body

Once a management system is fully implemented and stabilized, the measurement, assessment, and improvement of its efficiency becomes more important to contribute to overall organizational efficiency of the regulatory body.

Both self-assessment and independent assessment of the management system are expected to be conducted.

Suggested approaches

- It is recommended to establish processes for self-assessment and independent assessment of the management system. These processes have to cover all organizational levels and activities of the regulatory body and also define the frequency of such assessments.
- Different types of assessment can be used, for example, periodic management system review conducted by senior management, assessment of processes, internal audits, international peer reviews, etc.
- Assessment objectives are to identify gaps and weaknesses, identify potential enhancements in performance, and motivate the staff for continuous improvement.
- After the assessment is completed, it is recommended to develop a list of activities (e.g. action plan) designed to overcome gaps and opportunities for improvements identified in assessment activities and to monitor and report the status of actions.

4.4. KNOWLEDGE MANAGEMENT

The regulatory body of an embarking country may not have all required expertise in its own organization and may rely strongly on external support. However, the aim is that as the regulatory body matures, it builds competence to become self-sufficient. Knowledge transfer from the external support and knowledge management within the regulatory body are essential elements in the competence building of the regulatory body.

The same applies for the operating organization, which in the beginning may rely on the support of the vendor. The regulatory body, in its oversight activities, ensures the operator has an appropriate knowledge management system in place.

4.4.1. Generic challenges and approaches

Challenge No. 1: Management of tacit knowledge

While the management of explicit knowledge is relatively straightforward, the transfer and management of tacit knowledge is more challenging.

Suggested approach

- In transferring the tacit knowledge of experienced experts, different opportunities offered by the latest technology can be utilized; written reports are not the only way to

record or convey knowledge. A process to collect and transfer tacit knowledge will be created and included in the management system.
- The organization may encourage and enable sharing knowledge, for example, by arranging seminars, meetings, mentorship programmes and providing opportunities for the staff to discuss different matters and experiences.

Challenge No. 2: Transfer of tacit knowledge

The management system usually has some provisions for transfer of tacit knowledge of retiring experts. However, members of staff can leave for other reasons, and typically then there is only a short period of time available for knowledge transfer.

Suggested approach

- It could be beneficial if the management system has provisions also for these cases, not only for retirements. For example, the retrieving and recording of tacit knowledge may be a continuous process, not to be performed only when an expert is leaving the organization.
- The regulatory body may periodically conduct a competency need assessment which considers the vulnerability of the personnel knowledge loss risk.

Challenge No. 3: Contracts restricting knowledge transfer between organizations

Concerning knowledge transfer between different organizations, contracts can restrict the knowledge transfer, if the matter has not been considered when drafting the contract (e.g. between regulatory body and support organization, or between the operator and the vendor).

Suggested approach

- It is necessary to pay attention that the necessary clauses concerning the rights for the used or created information are included in the contract. Experts in contract law could be used when drafting or reviewing the contracts.

4.4.2. Phases 2 and 3 specific challenges

The generic challenges described above apply also to phases 2 and 3. Some specific features of the phases may intensify the challenges. These features and additional challenges are explained below.

Challenge No. 1: Knowledge transfer from external sources

The newly established regulatory body may not have all the needed expertise in its own organization, but it may utilize outside support instead. The external sources can be for example:

- Technical support organizations: The embarking countries are normally expected to use TSOs in support of the regulatory functions during initial stages of regulatory review and assessment.
- Experts hired to the regulatory body for a limited period. The initial staff of the regulatory body may include experienced experts (from other fields of industry,

governmental offices or from abroad) to help the regulatory body to perform its functions during the first years or during the construction of the first unit. These experts may leave after the agreed period.

- The vendor organization: Although the vendor does not perform work for the regulatory body, it may help in educating the regulatory body's staff in the technical questions concerning the reactor.
- The regulatory body of the vendor country or regulatory bodies of countries licensing or operating a similar type of reactor.

Suggested approach

- The process to transfer knowledge may be slightly different depending on the source of the knowledge, and this may be taken into account when planning and implementing the knowledge management.
- Attention will be paid to knowledge transfer when drafting contracts with external parties (see generic challenge concerning the contracts). This item is highlighted in phases 2 and 3 due to the extensive utilization of outside support.

Challenge No. 2: Knowledge management within the regulatory body

The situation within a newly established regulatory body is very different from a more established organization. In the beginning, the number of staff is increasing continuously, and intensive training of all the staff is being conducted. The training may include sending members of the staff to the regulatory body of the vendor country or other regulatory bodies abroad whose expertise is well established and recognized, or to other relevant organizations. The regulatory body may also participate in international activities, which is a good way to gain knowledge about lessons learned and best practices from other countries.

The challenge is how to share the knowledge efficiently within the regulatory body.

Suggested approaches

- Availibility of sufficient number of qualified persons for performing regulatory function for the long term needs of organization is ensured through strong and sustainable management system.
- The use of succession planning and mentorship programmes, supported by available technology, to enable effective knowledge transfer.
- Different practices can be used, for example after participating in external training, the participant may give a summary of the most important lessons learned in his/her own organization.
- If an expert is sent to work in another organization (such as the regulatory body of thevendor country), the expert may report regularly to his/her own organization of the experiences and lessons learned; this can be either written short reports or virtual events.

4.4.3. Phase 3 and early phase 4 specific challenges

The generic challenges of knowledge management apply to phase 3 and early phase 4. However, these phases have some specific features that are considered when planning and implementing knowledge management.

Knowledge management has interfaces to information management and document management. Information and document management can be considered as prerequisites for knowledge management.

Challenge No. 1: Ensuring knowledge transfer in case of high turnover of temporary staff

The regulatory body may use external resources or hire temporary staff for some oversight tasks during the construction project, due to the high workload or lack of necessary competences. In addition, the turnover of staff can be high, as positions in the industry come available as the construction project proceeds. Some staff of the regulatory body may choose to seek employment in the industry.

Suggested approach

- The described situation is considered in the knowledge management process. Sufficient resources need to be allocated to implement the knowledge transfer process.

Challenge No. 02: Ensuring availability of information throughout the lifetime of the nuclear power plant

The lifetime of an NPP usually stretches over several decades. During such a long period, technologies of recording information may change drastically. However, information recorded, e.g. during construction and commissioning of the NPP, ought to be available throughout its whole lifetime.

Suggested approach

- The records need to be maintained, and attention has to be paid to the advances in technology. If necessary, the essential original data need to be transferred to a form that ensures its availability in the future.

Challenge No. 3: Capturing information from the commissioning and early operation of the nuclear power plant

Valuable experience about the characteristics and behaviour of the plant and its SSCs is gained during commissioning and early operation. Recording all essential information in a timely manner may be challenging, but it is important for the future safe and secure operation of the NPP. Recording and preserving the knowledge is the responsibility of the operating organization. However, the regulatory body may wish to record its own observations from these phases.

Construction and especially commissioning can be hectic phases, and when working under pressure to keep to the schedules, some issues may arise. For example, the following has occurred:

- In commissioning of a new NPP unit, it was necessary to repeat some commissioning tests, as the result report was not prepared in a reasonable time after completing the tests. The responsible commissioning engineer left the company, and even if the actual measurement data from the tests was recorded and filed in an appropriate way, it was not possible to complete the test report only based on the measurement data (as the summary report also describes the conduct of the tests or any difficulties encountered in running the tests and the solutions to overcome them).

Suggested approach

- Recording the data from the commissioning and early operation is the responsibility of the operating organization. The regulatory body ensures that the operating organization has appropriate means for recording and preserving the necessary data in a form that can be retrieved later in the lifetime of the NPP. In its oversight, the regulatory body may pay attention to possible delays in reporting commissioning tests or unexpected events.
- The regulatory body probably observes at least the most important commissioning tests and early operation; it is a good practice for the regulatory body to develop a method to collect and record the observations and experience of its staff from these oversight actions. The data could include also lessons learned about the regulatory practices during these activities, not only observations of the plant behaviour.

Challenge No. 4: Recording justification of decisions made during commissioning and early operation

During construction and commissioning, it is probable that modifications are made, for example to the design or to the way the plant is operated. It is important that the justification for decisions is recorded and filed in a legible, identifiable, and retrievable way. This concerns also regulatory decisions.

The importance of records management is further emphasized in the case that the construction project has significant delays, and staff (both in the regulatory body and licensee and vendor organizations) will change during the project. While it may be seen as a trivial requirement that essential information has to be recorded and filed in a way it can be retrieved later, some problems may appear. Especially during construction and commissioning, with several parties involved and working under pressure to keep the time schedule, some issues may arise. For example, the following has occurred:

- Justification and reasoning behind some design solutions were not recorded, only the final solution. Later, modifications were planned when modernizing the plant. There were some peculiar solutions in the original design that were not understood as their justification was missing. In such a case, if the design is modified, it is possible that some essential feature is neglected, and that the modification impairs safety or security. In any case, more analyses and assessment are needed than if the reasoning for the original design solution had been fully documented.
- In a project to build a new nuclear reactor, the regulatory body and the licensee agreed on some matters in meetings in the beginning of the project. However, the meeting minutes were either not written at all or not filed in a way they would have been available later. This caused inconvenience, loss of time and extra work later in the project, as the licensee and the regulatory body had different recollections of what had been agreed on.

Suggested approach

- The processes to record information and manage documents are kept in place both at the regulatory body's and operating organization's premises well in advance before starting NPP construction. All the decisions and their justifications need to be recorded. Justifications that may seem self-evident at the time of the decision making may not be so clear years later for different persons.
- Concerning meetings, it is a good practice to agree at the latest in the beginning of the meeting whose responsibility is to write the minutes of the meeting and how the minutes will be approved and agreed on.

4.4.4. Phase 4 (operation phase) specific challenges

Challenge No. 1: Transfer of nuclear power plant safety knowledge and history to new staff

Sometime after the commencement of NPP operation, knowledge management presents a different challenge — new staff, having no exposure to the prior phases, will have entered the regulatory body, and the experienced staff of earlier phases will have begun to retire. Effective regulation would require that the new staff be well versed with the design bases and safety aspects of the installation as well as its complete operational, modification and safety history to date. That is, the new staff requires solid familiarity with safety issues that arose in the previous phases to properly evaluate the current safety issues and performance of the installation. Developing such a capability and perspective among the new staff is one of the main challenges of phase 4.

Suggested approach

- The regulatory body may consider establishing an in-house training department to implement a systematic approach to training. The department could arrange for experienced personnel to impart technical training to new staff. Knowledge transfer this way would cover not only documented information but also tacit knowledge about the significance of major events, safety issues dealt with and lessons learned for the future.
- The regulatory body could implement a structured mentoring programme whereby new staff shadows the experienced staff on the job, reinforcing what was learned in the classroom with practical conduct of the regulatory business.

Challenge No. 2: Diminishing responsibility for knowledge management

Once the major phases (i.e. construction, commissioning, fuel loading, and early operation) have been completed and the phase-appropriate knowledge management activities been conducted adequately by the regulatory body, it would become doubly important to practice knowledge management tools and techniques so that fresh experiential knowledge would be suitably recorded, shared, stored, and secured at par with safety significant knowledge of the earlier phases. Keeping abreast with research and its applicability to safety issues; ensuring critical review and application of operating experience feedback (OEF); encouraging cross-functional working on safety significant issues; performing succession-planning for positions important to safety; establishing and overseeing communities of practice; identifying critical knowledge workers; developing job profiles to monitor staff competence — these and many more seemingly disparate areas all concern knowledge management but the regulatory body may not be prepared to view or deal them through a knowledge management lens.

The challenge in this phase is determining and implementing an adequate mix of systemic measures that embeds knowledge management in the core regulatory body functions without introducing unnecessary or distracting work overhead.

Suggested approach

- The regulatory body's business objectives (or vision / mission statements) could emphasize that the regulatory body is a knowledge based organization and its foremost business is to achieved excellence in competence and knowledge of the staff. The regulatory body embeds knowledge management in its integrated management system which clearly identifies knowledge management responsibilites of top, senior and line management. Knowledge management activities will be assigned as a generic task to all core-functioning departments and their perfromance evaluation may include assesment against knowledge management indicators.

4.5. EXTERNAL TECHNICAL SUPPORT ORGANIZATIONS

The regulatory body is required to have the competences to perform its functions. It may, however, be necessary for the regulatory body to use the services of external experts or a technical support organization. The need for external expert support may arise because of:

- The need to complement already existing internal resources to perform activities with respect to unanticipated applications received from the licensee;
- A need to build up specific internal competences;
- A specific project for which special additional competences are needed;
- A need for a second opinion;
- Permanent outsourcing of certain activities (e.g. complex, specialized or infrequent activities).

4.5.1. Generic challenges and suggested approaches

Challenge No. 1: Hiring of consultants and assessment of their technical expertise

A newly established regulatory body may not have expertise in hiring consultants or TSO for development of regulations, competency development of regulatory staff, review and assessment, inspection and enforcement, etc. in different phases of the regulatory oversight process. Similarly, assessment of technical competency of an individual expert can be a challenging task for a regulatory body.

Suggested approaches

- The regulatory body could hire experienced persons from the regulatory body of the vendor country or from other Member States as senior consultants to support in identifying the areas in which technical expertise are required in phase 2 to early phase 4. These consultants may also be utilized to support the hiring of a TSO.

Challenge No. 2: Conflict of interest

The situation where same consultants or TSO are providing technical support to both operating organization and regulatory body may lead to conflict of interest.

Suggested approaches

- The regulatory body is required to verify whether the organizational structure of the provider of external expert support and its internal procedures provides functional and personal separation to ensure effective independence between units carrying out work for the regulatory body and units carrying out similar work for a licensee. The links between such units be carefully monitored. If a provider of external expert support is not entirely free from potential conflicts of interest, then an appropriate set of arrangement is established as a part of the contract with the TSO, for example not sharing the same resources, firewall protecting information, and non-disclosure agreements.

Challenge No. 3: Evaluation of work performed by experts

Evaluation of the work done by external experts or a TSO, and its use for decision making is a challenging task for the regulatory body at the early stage due to lack of required expertise.

Suggested approaches

- It may be difficult for the regulatory body to evaluate the work performed in technical areas due to lack of technical expertise. Therefore, the recommendations given by external experts or TSO may be utilized as one of the major inputs in regulatory decision making. However, the regulatory body is required to document the decisions it has made on the basis of input from the provider of external expert support. The basis for the decisions may also be recorded and documented appropriately for future reference.

Challenge No. 4: Confidentiality of information

The organization providing external expert support may have to address confidential information (security or protected information and proprietary information.)

Suggested approaches

- The external TSO is required to demonstrate that the access to such information is effectively restricted to individuals whose trustworthiness has been verified through a trustworthiness assessment (e.g. identity verification, criminal history review, polygraph testing etc.), and who have a need-to-know the information, that the information is kept under secure conditions, and that secure procedures are in place for communicating the information (secure communication channels, encryption capabilities, etc.), specific to the confidentiality level of the information.

Challenge No. 5: Non-availability of technical support from vendor country

Involvement of the regulatory body from the vendor country for technical support in different phases of regulatory oversight is a challenging task. The non-availability of technical support from the regulatory body of the vendor country in performing regulatory oversight at the time when needed in different phases might be due to involvement of regulatory staff in their own regulatory activities in the country because of expansion of nuclear power programme and lack of technical staff to spare for assistance to other countries.

Suggested approaches

- The regulatory body may approach the regulatory body of the vendor country to establish bilateral cooperation agreements, especially early in phase 3 when the negotiation of the contract for construction of the NPP(s) started with the vendor country. In this way, the regulatory body of the vendor country would have sufficient time to manage the recruitment and training of additional staff to oversee the regulatory activities within and outside the country. If technical expertise will still not be available, the regulatory body needs to avail itself of the opportunity to train their staff in the vendor country by attachment in different areas of expertise. In parallel, bilateral arrangements may be made with countries having plants of similar reactor technology.

4.6. EMERGENCY PREPAREDNESS AND RESPONSE

Radiation risks are very low during normal NPP operation, and they are under control by means of systematic radiation protection activities. However, there is still a certain residual risk of very low probability that large accidents with possible harm to the population and environment may occur. Therefore, EPR arrangements need to be in place and exercised before any emergency takes place.

4.6.1. Generic challenges and suggested approaches

Challenge No. 1: Advisory role and expert services of regulatory body in emergency preparedness and response infrastructure

The response to a nuclear or radiological emergency may involve many national organizations (e.g. the operating organization and response organizations at the local, national, and regional levels) as well as international organizations. The functions of many of these organizations may be the same for the response to a nuclear or radiological emergency as for the response to a conventional emergency. However, the response to a nuclear or radiological emergency or to a nuclear security event require expertise specific for nuclear power programme which is usually assured by the regulatory body.

The regulatory body may have an advisory role and strong educatory role for the overall EPR infrastructure due to its knowledge of nuclear and radiological risks.

The regulatory body may also provide expert services like radiation monitoring and risk assessment for actual and expected future radiation risks.

Suggested approaches

- It is good practice to develop basic expertise of the regulatory body for nuclear emergencies in early phase 2 and then to enhance it during phase 3; this expertise covers a wide range of technical and other knowledge, such as radiation protection, emergency exposure situations, nuclear technology matters, medical factors, etc.

- Communication capabilities of staff assuring the mentioned advisory role and expert services are crucial; expert knowledge ought to be transferred in a form understandable to staff of all participating organizations.
- It is common that the regulatory body needs support from a TSO for the most specialized activities like hazard assessment, etc.

4.6.2. Phase 2 specific challenges

Challenge No. 1: Functional coordinating mechanism for emergency preparedness and response

An EPR infrastructure exists in Member States before launching a nuclear power programme. This EPR infrastructure has to be strengthened based on the probability and severity of possible emergencies, both safety and security related.

Suggested approaches

- It is recommended to establish a coordination body in early phase 2 to coordinate with already existing or newly established bodies (e.g. a committee consisting of representatives from different organizations and bodies).
- Development of EPR infrastructure for nuclear emergencies can be managed as a project with defined activities, responsibilities, due dates, reporting mechanisms, etc. using project management tools. An action plan or any other plan can be used as well.

4.6.3. Phase 3 specific challenges

Challenge No. 1: Preparing and implementing practical communication arrangements in nuclear or radiological emergencies or security events

Effective EPR arrangements are put in place during phase 3. One of the key success factors of those arrangements is effective communication, which helps to maintain public trust in response organizations. In the preparedness phase, success in risk communication comes from good planning, good timing and the involvement of all interested parties. A communication plan is prepared not only by the operating organization or the regulatory body, but with the involvement of all actors.

Suggested approaches

- It is recommended that communication protocols are established between different organizations to ensure their clear roles, responsibilities and authorities.
- Proper communication campaigns and training about radiation risks can reduce anxiety. Informing and educating the public, authorities and others involved in the response to radiological health hazards is needed and need to be done systematically with a long term perspective.
- The communication language is important, so communicators are required to be trained and prepared to use simple language and correct terminology; use strong and simple visuals; to present scientific facts; to use active voice and personal pronouns; plain language background materials can be prepared jointly by communicators and scientists, particularly radiation experts, in advance.

- It is recommended to establish local and national processes and procedures for effective communication with the public during nuclear or radiological emergencies.
- It is recommended to coordinate EPR experts and public information professionals starting in the preparedness phase and particularly when designing preparedness tools, such as plain language briefing packages. This could be achieved through training and teamwork.
- Emergency exercises play a key role in strengthening the practice of public communication, and it is important to gather feedback from all interested parties. Emergency exercises can help to develop the relationship between EPR experts and public information professionals.
- Emergency exercises provide time for learning and the focus is not on whether an organization "fails". There is always room for improvement and once an exercise is successful, it is time to increase the challenge level to keep learning new things.
- Incorporating innovative media in communication arrangements is needed; social media play an important role in delivering information at a local and other levels.

4.6.4. Phase 4 specific challenges

There are no specific challenges for this phase beyond the ones previously identified.

4.7. INTERFACES, COMMUNICATIONS AND INVOLVEMENT OF INTERESTED PARTIES AND LICENSEE(S)

4.7.1. Generic challenges and approaches

Challenge No. 1: Developing effective engagement with interested parties

The national regulatory body in a country embarking on a new nuclear programme will have a wide range of interested parties with differing needs and expectations for communication and engagement. In the very early phase of the nuclear power programme, the government or the nuclear energy programme implementing organization will have the responsibility for the engagement with interested parties. However, once the regulatory body is launched as an independent entity, it can identify its interested parties and start to build relationships in order to earn trust and appropriately engage interested parties in its decision making. This is a challenge for the new regulatory organization.

Suggested approaches

- Development and implementation of a communication strategy integrated within the overall strategy of the regulatory body, and which has the visible support of senior leadership is an important first step. The regulatory body normally employs people possessing the relevant competencies to develop and implement its communication strategy.
- Within its communication strategy, the regulatory body needs to identify the full range of interested parties. The needs of interested parties range between the need for information only and the need for active consultation and participation in the decision making process. A map of interested parties, in which areas of common interest are grouped, may assist the regulatory body in developing proportionate interactions with each interested party.

- A stakeholder map would be underpinned by communication plans that may include the following elements:

  - Who is the stakeholder;
  - Their area of interest;
  - The regulatory body's key messages;
  - Who is the responsible contact or spokesperson;
  - How it will be delivered (media / conference / public forum / face to face meetings / website / social media, etc);
  - When the messages will be delivered (timing).

- The regulatory body also requires, through regulations or other means, its licensees to undertake communication and consultation with interested parties in relation to their licensed activities.

Challenge No. 2: Maintaining public trust

Developing and maintaining public trust is an essential role of a regulatory body. For any process of participation, there needs to be a certain degree of trust among all parties. If any interested party does not trust the regulatory body in a particular process setting, the legitimacy of the process might be weakened. Trust, once gained, is easy to lose and it needs to be earned on a continuous basis.

Suggested approaches

- Trust in the regulatory process can be enhanced by the public perception that the regulatory body is competent in its fields of expertise, is objective, reliable, transparent and responsive, and behaves fairly in interactions with interested parties. The regulatory body can make consultation with interested parties an integral part of the regulatory process, and regard interested parties as an asset that can contribute knowledge to that process and enable well informed decisions to be made.

Challenge No. 3: Effective communications between the regulatory body and the licensee

The licensee (or 'authorized party') is, obviously, a key player in the regulatory process. To fulfil its functions and responsibilities, effective two way communication between the regulatory body and the licensee(s) is vital.

Suggested approaches

- Communication expectations with the licensee(s) are to be set out in the regulatory body's policies and procedures.

  This could include a hierarchy of engagements, structured around strategic issues at a high level involving the most senior staff, which are underpinned by tactical and operational discussions. Beyond this lie the normal compliance and reporting engagements.

- The regulatory body at all times cultivates a relationship of mutual understanding and respect with its licensees based on frank, open and yet formal interactions, and maintain a constructive liaison on safety and security related issues.

Challenge No. 4: Making information available to interested parties

Over time, a trend has developed toward increased awareness of the need for transparency and openness regarding nuclear and radiation safety. Many countries have enacted laws related to the freedom of information which give the public the right of access to recorded official information. Yet, some sensitive information cannot be disclosed. Practices differ from State to State, depending on culture, history, government philosophy, legal and organizational factors. Thus, the regulatory body in an embarking country faces the challenge of striking the right balance between openness and protection of information related to its work.

Suggested approaches

- The regulatory body, within its national context, makes as much information as possible available to interested parties, including legal and regulatory requirements; conclusions from reviews and assessments; findings of inspections; and regulatory decisions. The regulatory body should also inform interested parties about its strategy, policies, procedures and management system.
- If the regulatory body provides general information to the extent possible, and explains the reasons for withholding any details on the basis of national criteria, interested parties usually will understand the need for such restrictions, as long as such rules are applied properly and are not abused.
- The regulatory body establishes requirements for protecting sensitive information, which may include nuclear security, proprietary information, or personal data, including procedures for trustworthiness checks of the individuals who have access to sensitive information.
- Each competent authority working in nuclear security also establishes and implement policy and procedures for the protection of sensitive information and sensitive information systems, including the appropriate sharing of information with other relevant agencies both nationally and internationally.

Challenge No. 5: Ensuring the availability of threats related information for interested parties

Usually, a large number of competent authorities is involved in the threat assessment procedure, including law enforcement agencies, intelligence services, regulatory bodies, defence forces, etc. The regulatory body also has an important coordination role in this process. Moreover, the application of the 'security by design' principle can considerably reduce the cost for the design of a physical protection system and, additionally, can increase the effectiveness of the design, but the relevant information has to be distributed among all interested parties. Therefore, ensuring the availability of all threat related information for all interested parties taking part in the threat assessment procedure and/or in the design of the physical protection system, is crucial.

Suggested approaches

- The regulatory body may ensure that, during phase 2, the operator and all relevant interested parties receive information regarding the DBT, which serves as a basis for planning the physical protection system of the NPP and for developing nuclear security measures for each stage of the lifetime of an NPP.

4.8. INTERNATIONAL COOPERATION

As explained in Section 3, international agreements bring some obligations to the regulatory body, but international cooperation also provides the regulatory body of an embarking country with an opportunity to benefit from the experience and knowledge of other organizations. Cooperation with the regulatory bodies of other countries can provide invaluable information and support.

4.8.1. Generic challenges and approaches

Challenge No. 1: Insufficient resources of the regulatory body

The resource planning may consider the international cooperation at an early stage to benefit from the resources available internationally. It is rather typical that, in the resource planning, the focus is more on the oversight tasks and in the development of regulations and guides. If not planned in advance, it may be difficult to find the necessary resources to fully benefit from the opportunities offered by international cooperation.

Suggested approach

- Participation in international activities has to be considered in resource planning. The government has to be informed about the needs of the regulatory body, as the government has the responsibility to provide the regulatory body with the competence and the resources needed. The government also has the responsibility of establishing a national policy and strategy for safety. The national safety policy and the strategy should take full benefit of effective participation in the global nuclear safety regime. The importance of participation in the global safety regime can be explained and highlighted in the discussions with the government, and in discussions about the regulatory body's budget.
- Growing use of communication technology can enable virtual participation and thereby reduce the time commitments and travel costs.

Challenge No. 2: Regulatory body of vendor country not having resources for cooperation (or not interested)

There can be several reasons for the lack of sufficient resources of the vendor country's regulatory body. They might already be supporting several other countries, or the oversight of the domestic fleet may require extensive resources. It can also be that the requests for cooperation from other countries were not considered during resource planning.

Suggested approach

- It is advisable to start discussing the cooperation as early as possible when the operating organization is making the contract with the vendor, so the regulatory body of the vendor country can take the work into account in its resource planning.
- If the vendor country's regulatory body, however, does not have the necessary resources, cooperation can still be sought out but perhaps with a more limited scope. Also, cooperation can be sought with other countries interested in or already constructing or operating a similar reactor.
- If the vendor country is not interested in cooperation in general, there is not much that the regulatory body can do. In that case, seeking out cooperation with other countries constructing or operating a similar type of reactor may be the best alternative and it is in any case recommended.

Challenge No. 3: Lack of common language

It is probable that the embarking country and the vendor country do not share the same language. This can lead to misunderstandings with the verbal and written information, ultimately leading to safety or security concerns.

Suggested approach

- As constructing and operating the reactor is a long term project, it might be worthwhile to invest in learning the language of the vendor country. Interpreters and translators can be used, but it is very valuable to have at least few persons as staff members with a sufficient skill of the language of the vendor country.
- The language of the project has to be defined in contractual documents at the start of the project.

Challenge No. 4: Dissimilarities in national requirements and regulatory approach

It may be that there are some differences in the regulatory approach or in the safety and security requirements between the embarking country and the vendor country.

Suggested approach

- Cooperation other regulatory bodies is still beneficial, even if some differences in the regulatory approaches exist. Recognizing the differences facilitates the cooperation and helps to focus on common themes.

REFERENCES

[1] INTERNATIONAL ATOMIC ENERGY AGENCY, Governmental, Legal and Regulatory Framework for Safety, IAEA Safety Standards Series No. GSR Part 1 (Rev. 1), IAEA, Vienna (2016).

[2] INTERNATIONAL ATOMIC ENERGY AGENCY, Security of Nuclear Material in Transport, IAEA Nuclear Security Series No. 26-G, IAEA, Vienna (2015).

[3] INTERNATIONAL ATOMIC ENERGY AGENCY, Nuclear Security Recommendations on Physical Protection of Nuclear Material and Nuclear Facilities (INFCIRC/225/Revision 5), IAEA Nuclear Security Series No. 13, IAEA, Vienna (2011).

[4] INTERNATIONAL ATOMIC ENERGY AGENCY, Establishing the Safety Infrastructure for a Nuclear Power Programme, IAEA Safety Standards Series No. SSG-16 (Rev. 1), IAEA, Vienna (2020).

[5] INTERNATIONAL NUCLEAR SAFETY ADVISORY GROUP, Nuclear Safety Infrastructure for a National Nuclear Power Programme Supported by the IAEA Fundamental Safety Principles, INSAG Series No. 22, IAEA, Vienna (2008).

[6] INTERNATIONAL ATOMIC ENERGY AGENCY, Experiences of Member States in Building a Regulatory Framework for the Oversight of New Nuclear Power Plants: Country Case Studies, IAEA-TECDOC-1948, IAEA, Vienna (2021).

[7] EUROPEAN ATOMIC ENERGY COMMUNITY, FOOD AND AGRICULTURE ORGANIZATION OF THE UNITED NATIONS, INTERNATIONAL ATOMIC ENERGY AGENCY, INTERNATIONAL LABOUR ORGANIZATION, INTERNATIONAL MARITIME ORGANIZATION, OECD NUCLEAR ENERGY AGENCY, PAN AMERICAN HEALTH ORGANIZATION, UNITED NATIONS ENVIRONMENT PROGRAMME, WORLD HEALTH ORGANIZATION, Fundamental Safety Principles, IAEA Safety Standards Series No. SF-1, IAEA, Vienna (2006).

[8] INTERNATIONAL ATOMIC ENERGY AGENCY, Leadership and Management for Safety, IAEA Safety Standards Series No. GSR Part 2, IAEA, Vienna (2016).

[9] INTERNATIONAL ATOMIC ENERGY AGENCY, Organization, Management and Staffing of the Regulatory Body for Safety, IAEA Safety Standards Series No. GSG-12, IAEA, Vienna (2018).

[10] INTERNATIONAL ATOMIC ENERGY AGENCY, Functions and Processes of the Regulatory Body for Safety, IAEA Safety Standards Series No. GSG-13, IAEA, Vienna (2018).

[11] INTERNATIONAL ATOMIC ENERGY AGENCY, The Fukushima Daiichi Accident, Report by the Director General, IAEA, Vienna (2015).

[12] INTERNATIONAL ATOMIC ENERGY AGENCY, Strengthening Nuclear Regulatory Effectiveness in the Light of the Accident at the Fukushima Daiichi Nuclear Power Plant, IAEA, Vienna (2013).

[13] INTERNATIONAL ATOMIC ENERGY AGENCY, Objective and Essential Elements of a State's Nuclear Security Regime, IAEA Nuclear Security Series No. 20, IAEA, Vienna (2013).

[14] INTERNATIONAL ATOMIC ENERGY AGENCY, Nuclear Security Recommendations on Physical Protection of Nuclear Material and Nuclear Facilities (INFCIRC/225/Revision 5), IAEA Nuclear Security Series No. 13, IAEA, Vienna (2011).

[15] INTERNATIONAL ATOMIC ENERGY AGENCY, Establishing the Nuclear Security Infrastructure for a Nuclear Power Programme, IAEA Nuclear Security Series No. 19, IAEA, Vienna (2013).

[16] INTERNATIONAL ATOMIC ENERGY AGENCY, Physical Protection of Nuclear Material and Nuclear Facilities (Implementation of INFCIRC/225/Revision 5), IAEA Nuclear Security Series No. 27-G, IAEA, Vienna (2018).

[17] INTERNATIONAL ATOMIC ENERGY AGENCY, Sustaining a Nuclear Security Regime, IAEA Nuclear Security Series No. 30-G, IAEA, Vienna (2018).

[18] INTERNATIONAL ATOMIC ENERGY AGENCY, Security during the Lifetime of a Nuclear Facility, IAEA Nuclear Security Series No. 35-G, IAEA, Vienna (2019).

[19] INTERNATIONAL ATOMIC ENERGY AGENCY, Nuclear Security Recommendations on Radioactive Material and Associated Facilities, IAEA Nuclear Security Series No. 14, IAEA, Vienna (2011).

[20] INTERNATIONAL ATOMIC ENERGY AGENCY, Nuclear Security Recommendations on Nuclear and Other Radioactive Material out of Regulatory Control, IAEA Nuclear Security Series No. 15, IAEA, Vienna (2011).

[21] INTERNATIONAL ATOMIC ENERGY AGENCY, Technical and Scientific Support Organizations Providing Support to Regulatory Functions, IAEA-TECDOC-1835, IAEA, Vienna (2018).

[22] INTERNATIONAL ATOMIC ENERGY AGENCY, Security of Nuclear Information, IAEA Nuclear Security Series No. 23-G, IAEA, Vienna (2015).

[23] INTERNATIONAL ATOMIC ENERGY AGENCY, Computer Security for Nuclear Security, IAEA Nuclear Security Series No. 42-G, IAEA, Vienna (2021).

[24] INTERNATIONAL ATOMIC ENERGY AGENCY, Computer Security Techniques for Nuclear Facilities, IAEA Nuclear Security Series No. 17-T (Rev. 1), IAEA, Vienna (2021).

[25]   INTERNATIONAL ATOMIC ENERGY AGENCY, Milestones in the Development of a National Infrastructure for Nuclear Power, IAEA Nuclear Energy Series No. NG-G-3.1 (Rev. 1), IAEA, Vienna (2015).

[26]   Convention on Nuclear Safety, INFCIRC/449, IAEA, Vienna (1994).

[27]   INTERNATIONAL NUCLEAR SAFETY ADVISORY GROUP, Independence in Regulatory Decision Making, INSAG Series No. 17, IAEA, Vienna (2003).

[28]   INTERNATIONAL ATOMIC ENERGY AGENCY, Managing Regulatory Body Competence, Safety Reports Series No. 79, IAEA, Vienna (2013).

[29]   INTERNATIONAL ATOMIC ENERGY AGENCY, Licensing Process for Nuclear Installations, IAEA Safety Standards Series No. SSG-12, IAEA, Vienna (2010).

[30]   INTERNATIONAL ATOMIC ENERGY AGENCY, IAEA Nuclear Safety and Security Glossary: Terminology Used in Nuclear Safety, Nuclear Security, Radiation Protection and Emergency Preparedness and Response, 2022 (Interim) Edition, IAEA, Vienna (2022).

[31]   FOOD AND AGRICULTURE ORGANIZATION OF THE UNITED NATIONS, INTERNATIONAL ATOMIC ENERGY AGENCY, INTERNATIONAL CIVIL AVIATION ORGANIZATION, INTERNATIONAL LABOUR ORGANIZATION, INTERNATIONAL MARITIME ORGANIZATION, INTERPOL, OECD NUCLEAR ENERGY AGENCY, PAN AMERICAN HEALTH ORGANIZATION, PREPARATORY COMMISSION FOR THE COMPREHENSIVE NUCLEAR-TEST-BAN TREATY ORGANIZATION, UNITED NATIONS ENVIRONMENT PROGRAMME, UNITED NATIONS OFFICE FOR THE COORDINATION OF HUMANITARIAN AFFAIRS, WORLD HEALTH ORGANIZATION, WORLD METEOROLOGICAL ORGANIZATION, Preparedness and Response for a Nuclear or Radiological Emergency, IAEA Safety Standards Series No. GSR Part 7, IAEA, Vienna (2015).

[32]   EUROPEAN COMMISSION, FOOD AND AGRICULTURE ORGANIZATION OF THE UNITED NATIONS, INTERNATIONAL ATOMIC ENERGY AGENCY, INTERNATIONAL LABOUR ORGANIZATION, OECD NUCLEAR ENERGY AGENCY, PAN AMERICAN HEALTH ORGANIZATION, UNITED NATIONS ENVIRONMENT PROGRAMME, WORLD HEALTH ORGANIZATION, Radiation Protection and Safety of Radiation Sources: International Basic Safety Standards, IAEA Safety Standards Series No. GSR Part 3, IAEA, Vienna (2014).

[33]   INTERNATIONAL ATOMIC ENERGY AGENCY, Communication and Consultation with Interested Parties by the Regulatory Body, IAEA Safety Standards Series No. GSG-6, IAEA, Vienna (2017).

ANNEX I.
CASE STUDIES FROM MEMBER STATES

I–1. FINLAND

I–1.1.   Nuclear power programme

There are two operating NPPs in Finland: the Loviisa and Olkiluoto plants. The Loviisa plant comprises of two pressurized water reactor (PWR) units of WWER type, operated by Fortum Power and Heat Oy (Fortum), and the Olkiluoto plant, operated by Teollisuuden Voima Oyj (TVO), comprises of two boiling water reactor (BWR) units and one new PWR unit (of EPR type) starting operation. At both sites, there are interim storages for spent fuel as well as final disposal facilities for low and intermediate level nuclear wastes.

Posiva, a joint company of Fortum and TVO, is constructing a spent nuclear fuel encapsulation plant and disposal facility. Posiva applied for operating licence for the facilities in the beginning of 2022.

Finland is currently reviewing a construction licence application for the Fennovoima Hanhikivi unit 1 (WWER type design) in Pyhäjoki. [1]

Furthermore, there is a TRIGA Mark II research reactor, FiR 1, in decommissioning phase.

I–1.2.   Regulatory body

The mission of the Radiation and Nuclear Safety Authority is "to protect people, society, environment, and future generations from harmful effects of radiation". STUK is an independent governmental organization for the regulatory control of radiation and nuclear safety as well as nuclear security and nuclear materials, as shown in Fig. I–1. STUK's tasks include participation in the processing of licence applications, supervision of compliance to the terms and conditions of the licence and carrying out inspections at the plants. Furthermore, STUK issues 'Regulations and Regulatory Guides on nuclear safety (YVL Guides)' and supervises compliance with them.

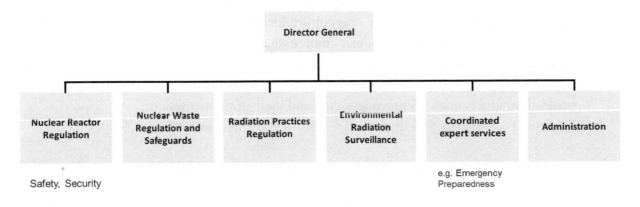

FIG. I–1. Organogram of STUK.

---

[1] Fennovoima has withdrawn its construction licence application in 2022.

59

I–1.3. Transition plan from oversight for construction and commissioning to oversight for operation

Regulatory requirements give the framework for the oversight. Important aspects in planning the transition to oversight the operation are:

- Construction and operation are very different phases, an important difference is that, during construction or pre-operational testing, mistakes do not cause immediate risk for nuclear and radiation safety and the mindset change from construction to operation does not happen overnight (concerns both the regulatory body and the licensee);
- Different competencies are needed for oversight of operation than for oversight of construction; thus, the oversight of operation may be performed by different experts than the oversight of construction. It is important to ensure continuity so that important lessons from construction are not lost;
- The commissioning phase is a unique learning opportunity also for the regulatory body, and staff is allowed time to observe commissioning tests;
- Pre-operational commissioning and preparation for fuel loading can be hectic and the time schedules can change constantly. The regulatory body may have flexibility in its resource planning and allocation to be able to react to the changes;
- The regulatory body starts oversight of licensee's preparations for operation early enough in the project.

I–1.3.1. Selecting important areas of oversight

For transition from phase 3 to phase 4, the focus is on ensuring preconditions of safe operation, such as:

- Competence and adequacy of staff (especially MCR operators);
- Leadership, management system and culture for safety;
- Procedures for normal operation and emergencies;
- Other documents and tools needed for operation (e.g. operational limits and conditions);
- Licensee's plans for radiation protection, waste management, maintenance, fire protection, emergency preparedness etc., and their implementation;
- Technical readiness of the plant and results of commissioning tests.

I–1.3.2. Feedback from review and assessment, and regulatory inspections

- The regulatory framework may allow flexibility to shift the focus of oversight based on observations and findings;
- The regulatory body may have practices to collect and analyze findings, observations and early indications of problems interdisciplinary; findings on one discipline may be an indication of problems within other discipline;

- Inspections are a good tool for verifying the licensee's performance, it is possible to focus inspections on areas where indications of possible problems exist.

I–1.3.3. Security and safety–security interface

STUK also oversees security in the use of nuclear energy. The requirements and guidance for safety and security are included in the same set of regulations and guides. Still, the safety–security interface is a challenge. Especially during incidents, other authorities (e.g. police, customs, boarder guard) have duties too, the roles and responsibilities have to be very clear. Security and the safety–security interface are considered during the entire lifetime of the plant. In operation, especially maintenance tasks, testing, modifications require well planned organizational and technical security measures. In safety–security critical organizations, the organization culture should be such as to consider safety and security matters in all decision making. Major lesson learned in these regards are:

- Access to security related information is restricted, but sufficient information flow ought to be ensured. For example, instrumentation and control has a significant interface with security.
- Clear rules and training for staff is needed. It is beneficial that all the staff, not only those working with security, are able to notice if security rules are breached.
- Strict key management may be in placed already during construction and pre-operational commissioning.

I–1.3.4. Lessons learned from the transition phase

Pressure to keep the time schedule may lead to proposals to postpone, closing some open items only later on, after operation has started. While in some cases this may be justified, it is not a good overall strategy. Also, the cumulative effects the remaining open items should be considered. If something is postponed, it needs to be communicated to all the parties that it is an exception. The regulatory body has to make its expectations clear to all parties early enough.

If the regulatory body has resident inspectors on the site, their role and rights are required to be well defined. If there are problems during commissioning, it is possible that the resident inspectors are approached and asked, for example, to permit to proceed despite the problems. The regulatory body needs to be prepared for that and have clear procedures for decision making in such situations.

Good safety culture is required to be maintained and enhanced during the whole project. It is very difficult to change attitudes and working methods quickly for start of operation. The licensee should be encouraged to take 'mental ownership' of the plant and of safety already in early stage of the project, even in a 'turnkey project'. This might be a challenge especially if the licensee is new with no previous experience of operating an NPP.

Sufficient communication between all parties is essential. The licensee, the plant contractor and the regulatory body should share a common view of the situation (for example on remaining significant open points before important milestones). Expectations are made clear; it is recommended that the regulatory body and the licensee (and plant contractors, when necessary) discuss well in advance for example the preconditions to proceed in commissioning and in the beginning of operation (fuel loading, first criticality, increasing power, etc.). It is good to be prepared for unexpected events in the beginning of operation (such as unexpected reactor trips during nuclear commissioning).

Procedures for making modifications during different phases (construction, commissioning, operation) have to be in place.

The commissioning phase is a unique learning opportunity (especially for the licensee, but also for the regulatory body), and it is ensured that the staff has sufficient time to participate and observe commissioning testing. End of construction and pre-operational testing pose some unique challenges:

- Several activities take place at the same time (finalizing installations, commissioning testing, operation), and they might have conflicting interests and involve different teams and responsible persons;
- Several organizations participate in the commissioning activities: the licensee, the plant contractor and, most important, the subcontractor. The roles and responsibilities (e.g. in the MCR) may be very different compared to operation phase;
- Some temporary solutions and procedures are probably in use.

It is ensured that duties and responsibilities, as well as how they change in the beginning of operation, are clear and which procedures cease to be valid when operation is started, and which new procedures come into force.

I–1.3.5.   Regulatory approaches to overcome the challenges

- Clear expectations and their communication to the licensee;
- Management system of the regulatory body: definition of duties, responsibilities and rights related to different positions, instructions for day-to-day work. The same is expected from the licensee;
- Establishing communication channels (both official and unofficial) with the licensee;
- Resource planning (covering both number of staff and needed competences);
- Use of support organizations, when needed;
- Important oversight tools for the transition phase:
    - ❖ Inspection programme (having flexibility on focusing inspections based on observations);
    - ❖ Observing activities on site (e.g. validation of emergency operating procedures, simulator training of MCR operators, implementation of security measures, emergency exercises, commissioning tests).

I–1.4.   Assessment and provision of competencies and resources

I–1.4.1.   Resource planning in STUK

The long term HR plan (5 years) mainly has statistical data on personnel, predictable retirements, needs for new resources (competence areas) and is updated annually. In addition, longer time scenarios (10–15 years) are considered.

The annual HR plan is done in every autumn. Section heads draft the list of tasks for the next year and estimate the number of resources and type of competence needed for carrying them out; this includes not only oversight tasks, but also development of regulations, participation in trainings (either as a pupil or a lecturer), etc. The estimation is compared with the existing resources and competences. In case of a gap, it is considered whether new persons are hired or whether the use of support organizations or consultants is possible (especially if the need of the resource or competence is only temporary). No special tools are used (Excel is used for listing the tasks and calculating the needed person days).

Finland has National Research Programmes on Nuclear Safety (SAFIR) and Waste Management (KYT) to develop and maintain the competencies in Nuclear Safety and Waste. All relevant interested parties participate (industry, universities, STUK, other governmental offices). Funding is based on payments from the licence holders (but other interested parties can fund projects they are interested in). Participating in the research projects is good education for young professionals.

The national training programme is a 6-week training course arranged every year. Around 70 participants every year and it covers all aspects of use of nuclear energy. All main interested parties participate in planning and implementing the course. The programme includes lectures, site visits, demonstrations with simulators, etc.

I–1.4.2. *Oversight of human and organizational factors*

Assessment of human and organizational factors (HOF) in STUK is an integral part of all oversights. Human and organizational matters have traditionally not been the core competence of regulatory authorities. However, HOF consideration is essential because, in nearly all incidents, there is a HOF in the background.

The regulatory body may have competence in HOF, but competence alone is not enough, when HOF is integrated into all oversight, and it may require internal training and changes in the working methods. HOF themes are included in all inspections. The inspections are planned together with the technical and HOF experts. There are HOF related internal trainings. STUK has an easy-to-use database where all the staff can record HOF observations and the HOF experts analyse the findings.

I–1.5. Communication with the interested parties

The interested parties are the utilities, industry, governmental offices, non-governmental organizations, and the general public. The regulations and regulatory guides have been translated into English and are available on STUK's website, this is important for new build projects as plant suppliers and major contractors are not Finnish. STUK organizes training about Finnish safety requirements both in Finnish (for licensees and Finnish contractors) and in English. When drafting new regulations or guides, the drafts are available in STUK's website for commenting (registration needed). Interaction with the licensees is essential. Interaction with the general public is also STUK's duty.

For oversight, it is essential to have good communication channels with the licensee, both formal and informal. STUK has:

- Established, formal channels exist for submitting applications, for regulatory decisions, for inspections and for informing STUK immediately about any major incidents;
- Informal communication (email, phone calls) when necessary (in practice continuously);
- Interaction with the plant supplier or contractors is always through the licensee;
- Regular meetings between STUK's and licensee's management;
- Regular meetings with the licensee in the new-build projects;
- Resident inspectors who can participate meetings on site (e.g. daily coordinating meetings).

I–1.6. Means of communication with the public during licensing

Regular meetings with the public living nearby the site. Regular meetings with the local media annually and when required. STUK gives presentation on the status of the regulatory reviews and results of its safety evaluations, answers questions, etc. STUK offers expert support when needed/asked by municipal board meetings, non-governmental organizations, and training courses for journalists concerning radiation related matters.

Results of oversight are published at STUK's website:

- Every four months: report of oversight activities, performed inspections and their results;
- Most important regulatory decisions;
- List of all applications from the licensees is also available (titles, not the content).

I–1.6.1. Meetings with the licensee and plant contractor in the Olkiluoto 3 project

During construction and commissioning, STUK had regular meetings with the licensee. The topics and intervals of the meetings varied according to the project phase and ongoing activities. Below are examples of regular meetings during the pre-operational commissioning phase:

- Management technical meeting (every second month);
- Commissioning meeting (every second month);
- Commissioning progress meeting (every week – teleconference);
- Fuel loading preparations (as needed);
- Operating licence meeting (four times a year);
- Instrumentation and control meeting (every week – teleconference);
- Project management meeting (four times a year);
- Resident inspectors can participate e.g. in daily morning meetings of the licensee and the plant contractor on site;
- Other meetings when necessary to discuss matters in more detail.

I–1.7. Lessons learned

It is important to write the minutes of the meetings, and to file them in a way they can be easily found later on. Since projects can last years (for example, the construction of Olkiluoto 3 started 2005 and first fuel loading was in 2021) and persons change, it is important to be able to trace earlier discussions and agreements. The role of the meetings needs to be made clear, are they only informative or are decisions made in the meetings. It is beneficial to share all meeting materials before the meeting to give all the parties the opportunity for internal discussion on the matters in advance.

I–2. HUNGARY

I–2.1. Facilities and activities

The nuclear fuel cycle in Hungary is open at both ends, since there is no production of nuclear fuels (all of them are being imported), and the spent fuel is not reprocessed but stored in an interim storage facility. There are four nuclear facilities in Hungary.

I–2.1.1. Paks Nuclear Power Plant

The Paks NPP is 110 km (southward) from Budapest and it is used for power production. It contains four PWRs of WWER-213/440 type, which were connected to the grid in 1982, 1984, 1986 and 1987 (construction 1975–1987), respectively. The fuel of the reactor is uranium dioxide, the fuel enrichment is 4.2% and 4.7%, respectively. Altogether, 349 assemblies can be inserted into the reactor core, and 312 out of these are fuel assemblies (37 control rods). The original electric power of the units was 440 MW, a power uprate to 500 MW was later carried out with the gradual efficiency upgrade of the traditional energetic components. The service lifetime extension of all units was approved by the Hungarian Atomic Energy Authority (HAEA) for additionally 20 years (expire date is 2037 for Unit 4 as the latest).

I–2.1.2. Spent Fuel Interim Storage Facility

The spent fuel assemblies of the Paks NPP power reactors are temporary (for 50 years) stored in the Spent Fuel Interim Storage Facility at Paks.

The Spent Fuel Interim Storage Facility is a modular vault dry storage facility with passive air cooling. The facility that is located just next to the site of the Paks NPP is operated by the Public Limited Company for Radioactive Waste Management (hereinafter referred to as PURAM).

I–2.1.3. Budapest Research Reactor (BRR)

The Hungarian built, Soviet designed 10 MW(th) research reactor is operated by the Hungarian Academy of Sciences Centre for Energy Research in the capital. It reached first criticality in 1959 (2 MW(th)). A power upgrade was carried out in 1967 (5 MW(th)) and total reconstruction was performed between 1986 and 1993 (10 MW(th)). The reactor is a tank type (water-water, beryllium reflector) reactor, the fuel is 19% enriched uranium (earlier it was 36%), and 229 fuel assemblies can be inserted into the core. The purpose of the research reactor is training, neutron physics, reactor physics research, and radioactive isotope production.

I–2.1.4. Training Reactor (BUT TR)

The pool type (water-water, graphite reflector) training reactor is in service since 1971 (construction 1969–1971) and is operating with a core built of EK-10 (uranium) fuel assemblies (23–24 assemblies in Al cladding) with 10% enrichment. The maximum thermal power is 100 kW. The reactor, located on the premises of the Budapest University of Technology and Economics, is operated by the Institute of Nuclear Techniques for research and training purposes (neutron physics, reactor physics research, instrumentation development, neutron activation analysis, and radiochemistry).

I–2.1.5. Mining property utilization company in the public interest

The Minewater Treatment Plant of the Mining Property Utilization Company in the Public Interest is operating in Kővágószőlős (southern part of Hungary, near to Pécs). In order to meet the health and environmental authority requirements, the uranium content of the surface and subsurface water contaminated by the former mining and milling activities is being removed in the Minewater Treatment Plant to protect the two water reservoirs located in the area. During a continuous process, the uranium is concentrated in the solid form of mixed uranium oxide ('yellow cake') that is exported (~2–3 tonnes/year).

I–2.1.6. Radioactive Waste Treatment and Disposal Facility

The Radioactive Waste Treatment and Disposal Facility at the Püspökszilágy site is operated by the Public Limited Company for Radioactive Waste Management (PURAM). It receives 10–20 m^3 low and intermediate level waste annually from smaller radioactive waste producers (hospitals, laboratories, and industrial companies), therefore several thousand disused radiation sources are stored there.

I–2.1.7. National Radioactive Waste Repository

The National Radioactive Waste Repository is located at the Bátaapáti site, it is for the disposal of the low and intermediate level waste originating from the Paks NPP.

I–2.1.8. Locations outside facilities with small amounts of nuclear materials

There are 43 licensed locations outside facilities having small amount of nuclear material. The 'small users' of nuclear sources are usually universities (Debrecen, Budapest, Veszprém), research institutes (ATOMKI, VEIKI, OSSKI) and other companies which use nuclear material for industrial applications.

I–2.2. Regulatory body

The HAEA's scope is the regulatory oversight of peaceful, safe and secure application of atomic energy in Hungary. Its scope comprises nuclear safety and security licensing, evaluation and oversight of nuclear facilities falling under the scope of the Act on Atomic Energy, the regulatory oversight of radioactive waste repositories, the registration and control of nuclear and other radioactive materials, the licensing of transportation and packaging thereof, and the regulatory control of nuclear and dual used items' export and import. Moreover, the scope includes evaluation and coordination of research and development, the performance of tasks related to nuclear emergency preparedness on the site, the approval of the emergency response plans of nuclear installations, and international relations. Licensing, inspection, reporting and oversight activities related to nuclear security and physical protection relevant to the use of atomic energy also fall within the competence of the HAEA.

In order to enhance the efficiency of the HAEA and to increase its independence, a legislative amendment was proposed and adopted that transformed the HAEA from an independent regulatory body into a public institution with special legal status. The change in legal status, which entered into force on 1 January 2022, also granted legislative powers to the president of the HAEA, allowing for a direct and efficient regulation in the field of nuclear energy.

I–2.3. Transition plan from oversight for construction and commissioning to oversight for operation

HAEA is responsible for the facility level nuclear safety, security and safeguards licensing of the construction and operation of the NPP, and also for SSC level permitting.

Facility level procedures follow the stages in the lifetime of a nuclear facility. Therefore, the site adequacy will be assessed keeping this in mind and before starting the essential part of the investment, technical plans, and safety analyses of the NPP will be fully developed, which the authority will review and, in case of their compliance, issue a construction licence. The constructed NPP will be put into operation, which means testing the functionality of implemented systems and then — after inserting fuel assemblies containing nuclear fuel — carrying out a series of measurements. If everything operates in accordance with the safety analyses and plans, and in compliance with the requirements, then the new reactor units can get an operating licence for a longer term. While conducting inspections HAEA verifies the compliance of the operation with the issued licences.

I–2.3.1. Site licence

During the site investigation and evaluation, the given site will be assessed in terms of whether it is suitable for accommodating an NPP. In addition to the above, all those characteristics of the site will be identified during its investigation and evaluation, which will be taken into account in the design of the NPP. The first procedure of the site licence is the regulatory approval of the site investigation and evaluation programme under a separate procedure in order to ensure that the specification of the site characteristics occurs in a systematic and planned manner, the programme covers all areas of expertise to be examined, and the suitability of the investigation and evaluation methods will be estimated before the start of the survey.

In the second procedure of the site licence, the application for a site licence can be submitted to the authority after the investigation and evaluations prescribed in the already approved programme have been executed. Detailed requirements for the site licence are contained in the Governmental Decree No.118/2011 (VII. 11.) Korm., Section 17, Paragraph 1, items a) and b) and Paragraph 19-a, as well as in the Nuclear Safety Code, items 1.2.2.0100 to 1.2.2.0800.
The site licence is effective until a maximum of 5 years from its issuance date and can be extended for additional 5 years. By granting the site licence, the nuclear safety authority accepts:

- The justification of lack of such site characteristics that would exclude the possibility of construction;
- The suitability of the conduction of site survey, assessment of data determined based on the site survey;
- The site related design data derived from the assessment and the suitability of the site.

I–2.3.2. Construction licence

It has to be demonstrated in the construction licence application that the NPP described in the licence application can be constructed and safely operated on the site provided with the site licence, the site properties specified during the site assessment and to be taken into account during the design have been fully taken into account and the NPP has appropriate protection against external hazard factors. A PSAR is attached to the licence application, in which it is demonstrated that the nuclear safety requirements for the NPP to be established that fall within the scope of the construction licensing procedure are met. The PSAR and the underlying documentation have sufficient level of detail that enables the authority to ascertain the fulfillment of the requirements without reviewing further documentation. Based on the application, the nuclear safety authority will determine the hold points to be used for control purpose and will plan the inspections.

The construction licence is effective until a maximum 10 years from its issuance date and can be extended for 5 years. Activities that can be performed on the basis of the construction license: preparation of the area necessary for the construction of an NPP, thus especially soil replacement and piling:

- Construction of the buildings and building structures of the NPP, construction of systems from system components with or without safety classification (manufacture, procurement and installation) according to the design furthermore, the appropriate construction of the entire NPP by the appropriate connection of systems. Performance of SSC function tests (without fuel assemblies containing nuclear material).

Detailed requirements in connection with the construction licensing procedure are prescribed in the Nuclear Safety Code, items 1.2.3.0100 to 1.2.3.0600. It is important to mention that the construction licence on its own does not permit the physical construction activities of the NPP. Additional SSC level permits (e.g. building, manufacturing, assembly, procurement) are necessary to start the actual construction work. Most of these permits — with a limited scope of exceptions, mostly related to site preparation — can only be acquired once the construction licence is issued. Also, based on the construction licence, a limited scope of plant commissioning is allowed without the use of nuclear fuel.

SSC level permitting is structured in a way to have a regulatory approval step at each significant design, manufacturing and installation stage. The following permitting steps are described in the Governmental Decree No.118/2011 (VII. 11.) Korm., and its appendices, the Nuclear Safety Codes:

- Manufacturing permit (chapter 1.3.2);
- Procurement permit (chapter 1.3.3);
- Assembly permit (chapter 1.3.4);
- Operation permit (chapter 1.3.5);
- Building or demolition permit (chapter 1.5.2);
- Occupancy permit (chapter 1.5.3);
- Construction or demolition permit for elevators (chapter 1.5.4);
- Occupancy permits for elevators (chapter 1.5.5).

Specific rules apply for each permit type on how and when they can be issued.

I–2.3.3.   Commissioning licence

In the licence application, it is to be demonstrated that the construction of the NPP is in compliance with the designer's intentions, and that the realized conditions are in compliance with the legal requirements. The licence application is accompanied by a preliminary version of the FSAR.

The commissioning licence is authorized to perform the following activities:

- The first placement of fuel assemblies into the nuclear reactor;
- Scheduled execution of commissioning programmes;
- Performance of tests of systems and system components important to nuclear safety under active circumstances;
- Operation at rated power in the case of NPP units from the date of the successful execution of the commissioning programme to the date specified in the licence.

The licence for commissioning is valid for 12 months from its issuance. Detailed requirements for commissioning licence are contained in the Nuclear Safety Code, items 1.2.4.0100 to 1.2.4.0600. The purpose of licensing nuclear safeguards is to prevent, detect and hinder the (further) nuclear proliferation, i.e. the non-peaceful uses of nuclear material. Prevention at the facility level can be implemented using so-called proliferation-resistant technical solutions that prevent nuclear proliferation, and through requiring rigorous inventory of nuclear materials, imposing reporting obligations and conducting official control, thereby the deterrence can be achieved due to a high probability of early detection of diversions and/or misuses. Basic legal requirements connected with the facility level safeguard system are considered yet at the designing stage, are summarized in the HAEA 'guide No. SG-2'. The application for first safeguard registration has to be submitted by the licensee at least 7 months before the arrival of fuel assemblies to the facility.

I–2.3.4. Operation licence

Necessary modifications during execution of commissioning programme are summarized and same is submitted as a modified FSAR along with the licence application. Further, it has to be demonstrated that the safe interim storage or final disposal of radioactive waste generated by the NPP including spent fuel, is in accordance with the latest proven scientific results and expectations and international experience.

In the possession of operation licence, the NPP can be operated under the conditions and for the duration specified in the licence. The nuclear safety authority sets the duration of the operational licence taking into consideration the operational features of the relevant nuclear facility and other circumstances, but it is not extended beyond the designed service life of the nuclear facility. Detailed requirements for the licensing of operation are contained in the Nuclear Safety Code, items 1.2.5.0100 to 1.2.5.0800.

I–2.3.5. Other oversight activities

To widen the scope of the regulatory oversight, in the licences HAEA may set hold points or witness points. Besides issuing licences, HAEA conducts regular inspections based on the annual inspection plan.

HAEA performs its assessment typically after termination of an activity or a process, so the conclusions are drawn subsequent to the completion of the activity or the process. The regular reports of the nuclear facilities are also source of the assessment.

I–2.4.   Communication with the interested parties

From 2013, the Act on Atomic Energy declares that, in all facility level licensing procedures, the HAEA arranges public hearings to ensure transparency and openness.

Before deciding, in order to obtain the opinion of the public, the HAEA holds a public hearing and informs the interested parties through public notice and by publication on its website, and the special authorities contributing to the procedure about the location and time of the public hearing in due time before the public hearing as determined in the Government decree.

In Hungary, the HAEA prepared a procedure for smoothly managing the organization of the public hearings. HAEA created rules to ensure equality and prevent long speeches and obstructive behaviour during the public hearings; these rules include inter alia:

- One question, or subject of opinion, written form, with name, organization name, or nickname;
- Three minutes timeframe for question or opinion;
- Three minutes timeframe for answers;
- The President appoints the person who will answer.

On the question form, besides the rules there is information about the final report, the video recording, and about how HAEA is going to utilize the outcome of the public hearing.

HAEA hires an independent professional moderator to make the audience keep the rules and to handle possible difficult situations. Every public hearing requires two summaries: HAEA's summary is about the licensing/decision making process, whereas licensee's summary is about technical information (what they plan, why, completed, and planned steps, benefits). To join the public hearings, no registration is necessary.

I–2.5.   Training

Training at HAEA is based on a systematic approach to training. The trainings include both civil service trainings and technical trainings. Civil service trainings comprise of basic training, special examination, and further training. Technical trainings have areas of inspectors training, Initial training, refreshing training and advanced courses. According to general procedure on training in HAEA, there are initial courses (quality assurance, informatics, personal protection, fire protection, basic professional procedures) and annual trainings.

The new employees need to complete an entry training. This entry training contains general and specific parts. The general part covers all tasks of the HAEA in general, while the specific part is knowledge about registry of radioactive materials, safeguards, physical protection, radiation protection, public administration, and legal knowledge. The staff also participate in national and international training courses organized by IAEA, European Commission, and United States Department of Energy.

I–3.   PAKISTAN

Pakistan has seven NPPs, of which five reactors are in operation, one is under commissioning, and one is in long shutdown phase for decommissioning. The Karachi NPP, unit 1 (K-1), was permanently shut down in August 2021 for decommissioning, after completing around 50 years of safe operation. Currently four NPPs are in operation at the Chashma site (designated as C-1 to C-4) and one at the Karachi site. The Karachi NPP unit 2 (K-2) started commercial operation in 2021, while first criticality of the Karachi NPP unit 3 (K-3) was achieved in February 2022. A summary of these power plants is given in Table I–1.

TABLE I–1. NUCLEAR POWER PLANTS IN PAKISTAN

No.	Plant	Status	Type	Capacity	Start of Operation
1	K-1	Permanent shutdown for decommissioning	PHWR	137 MWe	1972
2	C-1	Operation	PWR	325 MWe	2000
3	C-2	Operation	PWR	340 MWe	2011
4	C-3	Operation	PWR	340 MWe	2016
5	C-4	Operation	PWR	340 MWe	2017
6	K-2	Operation	PWR	1100 MWe	2021
7	K-3	Commissioning	PWR	1100 MWe	2022

PNRA has a vast experience of regulatory oversight of NPPs, research reactors, manufacturing facilities of safety class equipment, and radioactive waste management facilities. This case study provides guidance to embarking countries on the establishment of an independent regulatory body, competency developments of its technical staff and establishing technical support organization within the regulatory body.

I–3.1.   Establishment of an independent regulatory body

The history of use of nuclear power for electricity production dates back to 1965, when the Pakistan Atomic Energy Commission (PAEC) was established with the promulgation of the PAEC Ordinance as governmental and autonomous organization responsible for peaceful uses of atomic energy in the fields of agriculture, medicine and industry, generation of electric power, and performing associated research and development in Pakistan.

In 1984, a small Directorate by the name of Directorate of Nuclear Safety and Radiation Protection was created within the PAEC to look after the safety aspects of the nuclear facilities and activities in Pakistan. In 1990, the Pakistan Nuclear Safety and Radiation Protection

Regulations were promulgated which empowered DNSRP to perform regulatory activities for all nuclear installations of PAEC, and the use of ionizing radiation in agricultural, industrial, medical, and educational sectors throughout Pakistan.

Pakistan had progressively started working on the independence of the nuclear regulatory body in the early 1990s. However, signing the Convention on Nuclear Safety in 1994 pushed the Government towards independence of the nuclear regulatory body. As a transitory measure, the Pakistan Nuclear Regulatory Board (PNRB), within PAEC, was established to oversee the regulatory affairs till the establishment of a formal regulatory body. The Board was empowered to oversee the activities and performance of DNSRP and to approve regulations and guides. It consisted of a Chairman (Chairman PAEC), full time and part time members. The Director General of DNSRP was designated as one of the full time members and secretary of the Board.

In 2001, PNRA was established as a competent and independent regulatory authority for the regulation of nuclear safety, radiation protection, transport, and waste safety in Pakistan, and empowered it to determine the extent of civil liability for damage resulting from any nuclear incident.

I–3.1.1.   Formulation of organizational structure

The PNRA Ordinance has defined the initial structure of the organization according to which the authority will comprise of a chairman, two full time members and seven part time members having postgraduate qualification in nuclear sciences or nuclear engineering. relevant experience in radiation protection and nuclear safety is an essential requirement. The ordinance designated the Chairman as the Chief Executive Officer of the PNRA entrusted with day-to-day administrative responsibilities of the organization.

It also allowed creation of various Directorates to execute regulatory functions, enforce decisions of the Authority and supervise all matters related to nuclear safety and radiation protection. Accordingly, Chairman created various Directorates and was assigned with distinct tasks and functions. This structure comprised three technical directorates in Islamabad, namely the Directorate of Nuclear Safety (NSD), the Directorate of Radiation Safety (RSD), and the Directorate of Transport and Waste Safety (WSD). In addition, three Regional Nuclear Safety Directorates (RNSDs) were also established, which were designated as RNSD-I, RNSD-II and RNSD-III, in Islamabad, Kundian and Karachi, respectively. These directorates were functional under the executive wing and were reportable to member (Executive).

In July 2002, a Quality Management, Education and Training unit (QM&ET) was formed for initiating selection, recruitment and training of new officers and laying out the framework of a management system for PNRA.

The initial organizational structure did not have any organizational unit to look after the corporate affairs of the organization. The formal structure of a corporate wing of PNRA was raised in March 2004 when six new organizational units, including five directorates and one centre, were created. These directorates were named as Directorate of Regulatory Affairs (RAD), Directorate of Human Resource Development (HRD), Directorate of Policies and

Procedures (PPD), Directorate of Information Services (ISD), Directorate of International Affairs (IAD), and Directorate of Technical Support and Safety R&D (TSD). The organizational structure has been further updated based on the requirements over a period.

PNRA is the sole national regulatory body responsible for nuclear safety and radiation protection. Currently, the PNRA organization is divided into an executive wing and a corporate wing headed by member (Executive) and member (Corporate), respectively. The executive wing is responsible for executing the core processes, such as licensing and authorization and inspection and enforcement of the authority, whereas the corporate wing is responsible for review and assessment, safety analysis, regulatory framework, training and developing competence of staff and IT support. An internal TSO is also a part of this corporate wing, which provides technical support in licensing and regulatory decision making by the executive wing. The staff between both wings is rotated to familiarize with all core processes and utilize their experience and knowledge effectively.

The Secretary of the PNRA and the Director General of the Chairman Secretariat, report directly to the Chairman. The latter assists Chairman in coordination with other governmental agencies and in planning future activities of the PNRA. There are six Director Generals, two under member (Executive) and four under member (Corporate). They are responsible for execution of tasks and functions of Directorates and projects under their control.

I–3.1.2. *Enhancement of organizational capability*

In view of the 2005 energy plan of the Government of Pakistan for expansion of the domestic nuclear power programme, by constructing more NPPs to achieve an installed capacity of 8800 MWe by the year 2030, the regulatory responsibilities of PNRA were expected to increase significantly in the years ahead. The regulatory supervision of nuclear installations and facilities using radioactive sources required recruitment of more staff in the coming years and, consequently, their competency development in regulatory domains. PNRA launched the following Public Sector Development Projects to meet the current and future challenges. These projects enhanced the overall PNRA capabilities by expanding infrastructure and recruiting more staff:

- Institutional strengthening and capacity building of Pakistan Nuclear Regulatory Authority (PNRA) by means of the Centre for Nuclear Safety (CNS);
- PNRA School for Nuclear and Radiation Safety;
- Capacity Building of PNRA to implement a Nuclear Security Action Plan;
- Establishment of the National Dosimetry and Protection Level Calibration Laboratory;
- National Environnemental Radioactivity Surveillance Programme;
- Safety Analysis Centre (SAC);
- Further strengthening of PNRA capability for design assessment and analysis to ensure safety of NPPs in Pakistan;
- Establishment of the National Radiological Emergency Coordination Centre;
- Regulatory oversight against vulnerabilities of digital controls and cyber threats.

I–3.1.3. Staffing and competence development

PNRA was facing acute shortage of trained staff at the time of its creation. The problem was augmented by the fact that most of its available and much experienced workforce was retiring in near future. When PNRA was created in 2001, some of the existing staff from DNSRP was transferred to PNRA while some of its staff opted for the parent department, i.e. PAEC. The number of officers who joined PNRA was merely 38 at that time, which was clearly insufficient to cover the whole spectrum of activities related to regulatory business performing to existing and future nuclear and radiation facilities in the entire country. To cope with this challenge, PNRA decided to make fresh inductions immediately. In this regard, 'a two-pronged approach' was adopted. The first part of this approach covered a fast track direct recruitment drive, through which fresh university graduates from engineering and sciences disciplines and some experienced professionals from relevant industries were inducted. The second part was a relatively slow recruitment process, under which candidates were selected for fellowship scheme to complete, before joining PNRA, Master's degree programmes in Nuclear Engineering, System Engineering, Medical Physics, and Mechanical Engineering disciplines at the Pakistan Institute of Engineering and Applied Sciences (PIEAS), and Master's degree programme in Nuclear Power Engineering at the Karachi Institute of Nuclear Power Engineering (KINPOE). All the recruitment processes were carried out under the public sector development projects listed in Section I–2.1.2.

A systematic training need assessment of regulatory officials was performed in collaboration with the Lahore University of Management Sciences (LUMS) in 2002. This training need assessment was later recognized as a good practice by the International Regulatory Review Team (IRRT) Mission of the IAEA that visited Pakistan in December 2003. The training needs assessment utilized the so-called IAEA 'Four Quadrant Competency Model' for evaluation of needed competencies. Based on this competency model, a total of 52 training courses were identified for senior, intermediate and junior level PNRA staff working in various Directorates.

As already mentioned, the QM&ET unit was established in 2002 with the responsibility to induct new officers and arrange technical training for them. This unit started recruitment and first batch of young regulators joined PNRA in March 2003. A 12-weeks technical training course was arranged by QM&ET for this batch which was based on the IAEA's Basic Professional Training Course being held at Saclay, France. In September 2003, the first batch of candidates who completed their Master's degrees at PIEAS through fellowship scheme joined the PNRA. This batch also went the same training course. The QM&ET unit was later on, in 2005, transformed to the Directorate of Human Resource Development (HRD), which continued the induction with the support of the establishment branch.

I–3.1.4. Establishment of the PNRA School for Nuclear and Radiation Safety

To enhance, maintain and update the competency of PNRA employees inducted through the Public Sector Development Programme and to create a highly specialized and knowledgeable group of professionals in PNRA, an appropriate education and training of the regulatory staff was considered essential to meet the current and emerging staff demand for regulating nuclear and radiation facilities. It was necessary for each individual to be competent in their assigned tasks and maintain their competency throughout their professional career.

In 2006, PNRA submitted a proposal for the establishment of a training centre, namely the School for Nuclear and Radiation Safety under the Public Sector Development Programme. This proposal was approved by the Government which allocated sufficient financial resources for the establishment of the PNRA School. The main objective of the project was to produce professionals, technicians and managers skilled in nuclear and radiation safety working in PNRA, utilities and other interested parties.

I–3.1.5.   Establishment of the Nuclear Security Training Centre

The objective of the Nuclear Security Action Plan project was to develop a national sustainable system in nuclear security with the established emergency response and recovery capabilities, integrated with national laws, regulations and procedures to incorporate the gaps identified after having a made a political commitment with regard to the IAEA Code of Conduct on the Safety and Security of Radioactive Sources.

One of the focused areas of the Nuclear Security Action Plan was the Nuclear Security Training Centre (NSTC). The objective of establishment of NSTC was to develop a sustainable system of training in nuclear security with a comprehensive approach of prevention, detection of and response to incidents related to nuclear security. The interested parties of NSTC were regulators, operators, first responders, front line officers and law enforcing agencies.

I–3.1.6.   Establishment of the National Institute of Safety and Security

Both the School for Nuclear and Radiation Safety and NSTC had arranged a number of in-house professional training courses for technical officers of PNRA as well as its stockholders in the fields of nuclear and radiation safety, physical protection, and nuclear security. In 2014, PNRA merged these two training centres under one roof as a National Institute of Safety and Security (NISAS) for facilitating national and regional training courses on nuclear safety and security. NISAS is equipped with the classrooms and laboratories for training in the nuclear safety and nuclear security. NISAS has highly qualified permanent and visiting faculty members having Masters, PhDs degrees with post-Doctoral research expertise, and experience in regulating nuclear and radiation facilities.

Establishment of NISAS was a landmark towards its vision of becoming a global hub for high quality training in nuclear safety and nuclear security. NISAS is extremely confident to deliver training in all areas of nuclear safety and nuclear security to the satisfaction of its customers and foresee to get recognition at both national and international levels.

This institute is now a backbone of PNRA to provide sustainable technical workforce to carry out regulatory oversight activities at nuclear installation and radiation facilities for ensuring safety. The institute provides a highly effective learning environment and is capable of conducting 25 to 30 training courses each year, in which 500 to 600 staff from PNRA and other interested parties participate. This institute regularly hosts IAEA sponsored national, regional and international training courses and provides opportunities to fellows from foreign countries under the auspices of the IAEA.

The IRRS mission to Pakistan in 2014 recognized PNRA's well developed training programme to maintain competence of regulatory professionals for rapid expansion of PNRA as a good practice.

I–3.2. External technical support

PAEC started generation of electricity from NPPs with the commercial operation of first CANDU reactor in 1972. In 1991, PAEC planned to expand its nuclear power programme and signed an agreement with China National Nuclear Corporation (CNNC) for construction of a 325 MWe PWR nuclear power plant at the Chashma site, District Mianwali, Pakistan. The DNSRP of PAEC was responsible to perform regulatory activities for all nuclear installations. However, capabilities were very limited for site evaluation, review, and assessment of the PSAR, regulatory inspections during civil construction, installations and manufacturing of equipment, review and assessment of the FSAR and subsequent submissions.

Consequently, to cover these challenging tasks, DNSRP had signed a protocol with the National Nuclear Safety Administration (NNSA), the regulatory body of China, for cooperation in the field of nuclear safety. Under this protocol, an agreement was signed between DNSRP and the Nuclear Safety Centre (NSC) located in Beijing, which is a technical support organization of NNSA for providing assistance to DNSRP in the review of SARs and regulatory inspections during civil construction, installations, commissioning and manufacturing of equipment. A series of training courses were conducted by NNSA/NSC in Pakistan to train DNSRP personnel from 1991 to 1992. A group of DNSRP personnel were placed at NSC for three months in 1992 to understand the safety aspects of NPP design, SARs, codes and standards under the supervision of senior NSC personnel before participating in the review of the PSAR.

Review of the PSAR of Chashma NPP unit 1 (C-1) was jointly performed by NSC and DNSRP personnel in Pakistan. During this review process, junior officers of DNSRP participated in review activity as observers in order to get familiarization with the review process. Meanwhile, DNSRP started development of its own capabilities from IAEA and vendor country. In 1999, the FSAR was reviewed by DNSRP, and licensing queries were shared with NNSA for feedback. However, NSC personnel participated in the review meeting along with DNSRP personnel.

In 2001, PNRA was established as independent regulatory body and DNSRP was merged in PNRA. PNRA decided to gradually reduce the dependency on foreign expertise for licensing of upcoming NPP units by developing the required technical competence internally. This decision was made to maintain and demonstrate regulatory independence and for further strengthening of regulatory oversight capabilities to ensure safe use of nuclear power. Accordingly, in case of licensing of second PWR unit (C-2) later on in 2006, PNRA performed a review of the PSAR independently and got support, in few chapters only, from NSC which also participated in the review meeting.

In order to gain additional confidence, chapter wise IAEA expert mission were conducted under the technical cooperation project of IAEA which was titled "Application of New IAEA Nuclear Safety Standards for Licensing of Chashma Nuclear Power Plant unit 2 (C-2)". The objective was to evaluate the review performed by technical team of PNRA in the initial phase of independent review. During this review process, PNRA personnel interacted with international experts in different areas following which they jointly participated in the review meeting with the designer. As a result, the review team developed their confidence and identified additional areas of improvement including specialized areas of safety analysis.

Licensing of manufacturer of nuclear safety class 2 and 3 mechanical equipment was another area where external support from NNSA, China for review and assessment and inspections was needed. Heavy Mechanical Complex-3 (HMC-3) submitted its licence application in January 2004 for manufacturing of 'nuclear safety class 2 and 3 mechanical equipment'. As of that time, formal licensing process and regulatory requirements for nuclear equipment manufacturing were not devised. Henceforth, PNRA initiated its efforts for formulating provisional licensing requirements based on well recognized international licensing and certification practices and procedures. PNRA, with the assistance of NNSA (Chinese regulatory body), prepared a detailed process for review of technical details/documentation and evaluation of applicant's capability through rigorous inspections and established requirements in light of applicable international standards. After fulfilment of all the regulatory requirements, PNRA authorized HMC-3 to manufacture safety class 2 and 3 equipment for NPPs. The manufacturing licence was issued in 2005.

In order to further enhance the capability of PNRA personnel in pre-service inspection and in-service inspection, operating experience feedback, NPP design and ageing management, external support was needed. Therefore, PNRA signed a memorandum of understanding (MoU) with the Research Institute of Nuclear Power (RINPO) of China in June 2007 for cooperation and assistance in these areas. Later on, RINPO became the China Nuclear Power Operation Technology Corporation, Ltd (CNPO). A number of PNRA technical staff were trained at CNPO in different disciplines like pre-service inspection and in-service inspection, operational management OEF and information exchange, management techniques for NPP under construction, NPP design and ageing management. Several joint technical workshops in the areas of interest were also arranged at PNRA headquarters, Islamabad.

I–3.2.1.   *Establishment of a technical support organization*

The regulatory review and assessment and inspection of nuclear installations are amongst the main core functions of a regulatory body, which are required to be performed to assess safety demonstration and compliance of standards by the licensee during the licensing process. The regulatory professionals require comprehensive technical expertise in these areas. PNRA, in an early stage of its establishment, recognized the need for development of technical competence of its staff to gradually reduce the dependency on foreign expertise for licensing of upcoming NPPs. The aim was to maintain and demonstrate regulatory independence and further strengthen regulatory capabilities to ensure safe use of nuclear power. PNRA planned to establish a dedicated organizational unit which would provide technical support in its regulatory activities.

Accordingly, PNRA submitted a proposal to the Government of Pakistan under its Public Sector Development Programme for creating a Centre for Nuclear Safety (CNS) within PNRA as an internal technical support center. The main objectives of establishing the CNS were to strengthen and enhance the existing regulatory capabilities of PNRA to discharge its responsibilities regarding the licensing of the upcoming NPPs, developing strong documentation base; and developing and strengthening bilateral links with TSOs of other countries. The Government approved PNRA's proposal in October 2004 and the CNS was formally established in June 2005.

After its establishment, the CNS collaborated with NSC of China for trainings in areas of review and assessment and safety analysis. Similarly, the Regional Offices of PNRA were also trained in regulatory inspections. The CNS also benefitted a lot from IAEA technical cooperation projects by attaching technical staff to various institutes around the world to enhance their technical skills. It is well known that continuous training and retraining in specialized fields is an essential element for maintaining competency of the regulatory staff. In this context, TSO staff also participated in various trainings courses at NISAS, and some training courses were arranged at national organizations such as PIEAS. For capabilities related to welding and relevant codes, TSO staff was trained at Pakistan Welding Institute (PWI). Similarly, for level-1 and level-2 in non-destructive examinations, TSO personnel were trained at National Centre for Non-Destructive Testing (NCNDT). Various other trainings were imparted to TSO personnel at national institutes like Pakistan Standards and Quality Control Authority (PSQCA), Pakistan Institute of Management (PIM), and Pakistan Manpower Institute (PMI).

As a result, the TSO staff gained significant knowledge through participation in trainings both at national and international levels. Thanks to the participation of regulatory professionals in the review and assessment process, TSO staff was able, within a period of five years after its establishment, to carry out the actual task of review of SARs of NPPs indigenously, although the external Chinese support was available and minimally utilized in special/complex safety issues. In 2010, CNS reviewed the FSAR for C-2 independently. Similarly, review of SARs of Chashma NPP units 3 and 4 and Karachi NPP units 2 and 3 (K-2 and K-3) was performed by CNS. Supporting safety analysis and independent audit calculations in required areas of SAR such as deterministic and probabilistic safety analysis, stress and structural analysis were performed by TSO staff by using internationally well recognized computer codes (such as RELAP, MELCORE, ANSYS, LS-DYNA, and Risk Spectrum).

CNS has trained its staff in the areas of deterministic and probabilistic safety analysis. The analytical capabilities provided support to review and assessment process. These codes enabled CNS to perform audit calculations submitted by the licensees and perform research and development in these areas. In order to meet the challenges for grey areas of review and assessment and safety analysis, further collaboration with NSC and IAEA technical cooperation projects was explored. In order to reduce the communication gap between PNRA and NNSA, PNRA trained its staff in Chinese language and placed them at NNSA/NSC to work with Chinese counterparts and enhance understanding of relevant technical documents of the vendor country.

Apart from licensing review of NPPs, the TSO also performed review and assessment of the SAR of the spent fuel cask to be stored in the dry storage facility in the country. In addition, the TSO performed review and assessment of various other licensing submissions to support the regulatory decision making at different stages of the licensing process. These include the review of periodic safety review reports for licence renewal after every 10 years of NPP operation, the review of modifications in design and technical specifications, etc. With increase in construction of new NPPs in Pakistan, PAEC commenced manufacturing activities of safety class equipment. In the initial phase of assessment of licensee's submissions with regard to design and manufacturing of safety class equipment, PNRA obtained assistance of the Chinese regulatory authority (NNSA). However, the TSO has become capable to perform assessment of similar applications in future, and award licences.

In order to integrate NPP operations knowledge in regulatory decision making, selected technical staff of the TSO went through rigorous NPP operation training. These personnel went through whole trainings required for plant operating personnel and obtained the licences to operate NPPs. They remained involved in operations at NPPs for a certain period of time after completion of training and then resumed duties in the TSO. PNRA benefitted from their knowledge and experience of those TSO personnel who obtained power plant operation training in many ways. For example, they imparted the related knowledge to other personnel of TSOs performing review and assessment of SARs related to plant systems, technical specifications and review of plant programmes. These licensed operators are also involved in licensing examination of plant operating personnel. For understanding of industrial codes and standards, in-house studies for comparison of existing and old version of standards were performed.

OEF is a key element during all phases of NPPs. OEF is utilized in review of PSAR, FSAR and periodic safety review. A dedicated review period is allocated in review schedule to review the licensee submissions in the light of national and international experience feedback available through the IAEA and other forums and organizations. Consideration of OEF also enhanced understanding of TSO personal and resulted in building confidence of the TSO.

The establishment of the TSO has provided a pool of experts to support various regulatory assessments of siting, design, construction, commissioning, operation and review of design modifications of NPPs. This has contributed to providing an independent insight for ensuring safety of the NPPs. It has enabled the regulatory body in timely, cost effective and efficient regulatory decision making for new and existing NPPs. The TSO has enhanced the capability of the regulatory body for utilizing international OEF to enhance the safety of nuclear installation. Assessment of lessons learned from the Fukushima Daiichi accident is an example of utilizing international OEF. Interactions with designers and licensee enhanced the confidence and knowledge of reviewers.

The development of the TSO has enabled the regulatory body to develop indigenous capabilities in different technical areas. Within a period of 10 years, the regulatory body became self-sufficient in performing the core regulatory functions of review and assessment and regulatory inspections without any dependence on external expert's support. The growing role of the TSO signifies that PNRA has substantially decreased its reliance on experts of foreign countries in the assessment of NPP safety. Currently, the TSO is also looking forward to gain expertise in emerging technologies and other challenging areas.

I–3.3.   Conclusion

PNRA is the statutory, independent and sole authority to regulate all applications of nuclear energy in Pakistan. It is staffed with competent staff having diverse experience in all regulatory activities and processes pertaining to different phases of NPPs ranging from siting to decommissioning. PNRA officers are recognized experts in IAEA and are being involved as experts, lecturers, reviewers and consultant in various technical missions, training courses and workshops. PNRA has contributed to the development of training material for the IAEA for competency development of newly established regulatory bodies in embarking countries.

PNRA is capable to provide help to countries embarking on nuclear power programmes in all nuclear regulatory domains. PNRA can provide full spectrum of support to an embarking regulatory body for independently performing regulatory functions related to nuclear installations and radiation facilities. It can also provide assistance in safety review/assessments and inspections, necessary for authorization and licensing during the lifetime of an NPP.

ANNEX II.
RELEVANT GOOD PRACTICES IDENTIFIED BY IAEA PEER REVIEWS

This Annex highlights selected good practices identified during IAEA's Integrated Regulatory Review Service (IRRS) missions and International Physical Protection Advisory Service (IPPAS) missions that are relevant to the challenges and suggested approaches which may be helpful for the management of regulatory oversight for the operation of a first NPP.

"An IAEA Good Practice is identified in recognition of an outstanding organization, arrangement, programme or performance superior to those generally observed elsewhere. It will be worthy of the attention of other regulatory bodies as a model in the general drive for excellence." [II-1]

II–1.1.   Organization, staffing and competency

II–1.1.1.   Organizational structure of the regulatory body and allocation of resources

The regulatory bodies need to ensure they have an organizational structure which ensures that resources are effectively allocated to enable them to discharge their duties under the legal and regulatory framework of the State. Good practices as quoted below are beneficial in this regard[2].

a) "The Office for Nuclear Regulation (ONR) developed its matrix management structure that effectively allocates resources to need. It also improved its hiring, training and strategic planning practices so as to develop new hires and to effectively anticipate and fill future needs. ONR matched its resources to needs using a matrix structure that also involved a strategic look-ahead. The five regulatory Divisions of Office for Nuclear Regulation operated in a matrix management arrangement, whereby four divisions, known as 'delivery areas' form the columns, each with a delivery lead. The rows comprised specialist resources, all of which were functionally located in the Technical Division. There were approximately fifty technical areas, grouped into fifteen technical specialisms, each with a professional lead. Resourcing discussions between the delivery leads and professional leads were held on a regular basis to ensure that appropriate resources were applied to meet the needs of each delivery area." [II-2]

b) The competent authority appears to have adequate resources and government support to perform its legislative mandate to oversee an effective State system of physical protection.

c) The competent authority recognized the benefit for a close cooperation with other State organization to cross-study complex issues in depth. In the framework of their research activities for which a dedicated budget is yearly allocated, the competent authority set up a working group in coordination with the national defence university to assess and implement ways to deal with insider threat.

II–1.1.2.   Staffing and competence of the regulatory body

---

[2] The good practices presented at II–1.1.1 (b), II–1.1.1 (c), II–1.2.1(c), II–1.2.1(d) , II–1.2.1(e), II–1.3.2 (b), II–1.3.2 (c), II–1.3.2 (d), II–1.3.2 (e), II–1.4.1 (f), II–1.4.1 (g), II–1.6 (d) and II–1.6 (e), originate from IPPAS missions. The corresponding text in these bullets was customized for information purposes only, to avoid disclosing the identity of the regulatory bodies involved.

Establishing and maintaining the staff competence is of the challenges and fundamental expectation across the international nuclear community. Regulatory bodies have established different mechanisms to cope the challenge.

a) "In order to share the regulatory experience and enhance the regulatory effectiveness, the Ministry of Environment Protection (MEP) set up groups of different special areas using social software (QQ & WeChat), for example: in the area of nuclear technology utilization, there were 5 groups on QQ network, including one director group, two junior inspector groups, and two system manager groups, with more than 1000 users.

b) The regulatory body established a variety of tools to manage their rapid growth and adopted innovative approaches to building a healthy organizational culture. Innovative practices included delegating responsibility for preparing the knowledge management strategy to newer staff, holding day-long meetings with staff to solicit feedback and holding a competition for staff to prepare essays on potential improvements (and establishing working groups to implement these improvements).

c) The National Nuclear Regulator supported the recruitment of qualified and experienced persons to its vacant positions through a joint bursary and internship programme. To facilitate the recruitment of qualified persons having workplace expertise, the NNR supported bursary students in various fields of science and engineering at higher learning institutions and operated an internship programme for freshly graduated persons. The NNR had a Succession Planning Policy and Procedure in place that addressed the replacement of staff in critical positions, including retired persons. To facilitate the recruitment of qualified persons having workplace expertise, the NNR supported bursary students in various fields of science and engineering at higher learning institutions and operated an internship programme for freshly graduated persons.

d) The Australian Radiation Protection and Nuclear Safety Agency had a well-developed strategy to compensate for the departure of qualified staff that systematically assessed succession risks for every position in the organization and prioritised the development of competencies that were found to be vulnerabilities to the long-term capability of the organization. In order to identify potential future resource risks, ARPANSA had systematically assessed every position in the organization to identify knowledge management and succession risks and identify mitigation measures to address any risks. These measures had been prioritized based on the risk of losing an essential competency." [II-3]

II–1.2.   Core regulatory processes

II–1.2.1.   Regulations and guides

Taking international feedback into account, the proactive development of new regulations for small modular reactors (SMRs), development of guidance in case nuclear security events, and protection of IT systems are good practices recognized by IAEA peer review missions.

a) "The prompt and integrated approach of the Swedish Radiation Safety Authority (SSM) to establish a consistent and comprehensive regulation taking into account international standards and good practices. The comprehensiveness and the expediency by which the ongoing regulation update projects were carried out using an integrated approach and taking into account international standards and good practices were commended. Swedish Radiation Safety Authority presented a new structure for the regulations describing the three levels of regulatory control under the law and ordinances and explained that all the Swedish Radiation Safety Authority regulations would be included in that new structure.

Swedish Radiation Safety Authority had decided to create supporting documents describing the rationale behind the regulations and would include formal interpretations of the regulatory sections. Both projects used IAEA standards and good practices of other countries as input and would also be used to implement the WENRA Safety Reference Levels, the European Directive for Nuclear Safety and the Basic Safety Standards. The projects were expected to be completed by February 2018. The ongoing regulation update projects would provide for a consistent and comprehensive set of regulations that would enhance the stability and consistency of the Swedish regulatory framework." [II-4]

b) "The Canadian Nuclear Safety Commission proactively developed extensive guidance and processes to assist potential applicants determine the content of the SMR application. SMRs might have a significantly different demonstration of safety than existing reactors. CNSC provided guidance on pre-application opportunities to ensure vendors understand the regulatory requirements and to provide them with an appropriate application assessment strategy that included a risk-informed assessment of the Safety and Control Areas (SCAs) and the use of alternative approaches in the development of the licensing application." [II-5]

c) The regulatory body's provision of comprehensive advice and guidance to interested parties for the planning and preparedness for and in response to nuclear security events, resulting in a high level of preparedness at civil nuclear sites.

d) Although as legally non-binding instruments, the regulator's guidelines represent, according to their number, structure and content, a great tool for the users to comply with the legally binding requirements by showing them the way that is most advised by the regulator. The ongoing work on the preparation of the 'Physical protection requirements for new nuclear power plant units (PP-17)' and 'Protection of IT and ITC systems (PP-18)' shows orientation of the regulator to the future objectives, roles, and responsibilities.

e) The provision of a design basis threat to new nuclear facilities at the early phase of project implementation provides a potential licensee with the best possible opportunity to take advantage of 'security by design' and plan and implement nuclear security arrangements at early stages.

II–1.2.2.   Review and assessment

a) "Bel V, the technical arm of the Belgium regulatory body, had developed and implemented an effective tool, with well-defined criteria applying a graded approach for reviewing safety related modifications, termed 'non important modifications.' (...) A tool had been established with well-defined criteria to establish a clear graded approach. A scoring sheet had been developed, with two groups of criteria: importance for safety and complexity of the non-important modifications." [II-6]

II–1.2.3. Inspection

At Swiss Federal Nuclear Safety Inspectorate (ENSI), 'the safety culture focus groups' are an effective tool for proactively engaging the senior management of NPP operators to promote self-awareness of their impact as leaders on the safety culture of their organizations.

a) "Every three years the Swiss Federal Nuclear Safety Inspectorate's (ENSI) Human Factors Organization holds a dialogue on safety culture with the senior leadership teams and safety culture specialists of the NPP licensees. This focus group is a facilitated discussion of a topic related to safety culture, with the intention of promoting self-awareness among the senior leaders of their demonstrated attitudes towards safety culture and the resulting potential impacts to the members of the organizations they lead." [II-7]

II–1.3.   Management system

II–1.3.1. Management system of the regulatory body

a) "The ARPANSA applied a holistic and comprehensive way of integrating all types of risks in the management processes, the regulatory activities, and day-to-day work activities, providing a strong foundation for their performance management framework. ARPANSA's Risk Management Framework was the core of the integrated management system (IMS) while it was embracing a range of risks such as risks to the ability to carry out the statutory functions; risk to employees and assets; as well as radiation risks to people and the environment. This comprehensive way of integrating all types of risks set the foundation for how the performance management framework was built." [II-8]

II–1.3.2. Culture for safety

Below mentioned practices are a practical source of information for embarking countries to cultivate nuclear safety and nuclear security culture in their organizations.

a) "Safety culture was an integrated part of the management system and the Radiation Protection Inspectorate (RPI) assessed the technical staff's experience on the safety culture aspects including leadership for safety. The issue of safety culture was incorporated in the management system. RPI put in means for assessing the staff's experience on the safety culture aspects including leadership for safety and also an assessment of safety culture was performed in the technical staff.
The issue of safety culture was incorporated in the management system. RPI developed a tool to assess the safety culture on a yearly basis. The tool consisted of a questionnaire to the staff about their experience on the level of safety culture including questions on leadership and of safety culture. At the daily meetings in the different Divisions, the staff are encouraged to have a questioning and learning attitude supported by the managers, as part of a contribution to fostering and sustaining a strong safety culture." [II-9]

b) The competent authority, supported by the TSO and the nuclear operators, have taken significant measures to build and begin to implement a comprehensive programme to enhance the security culture of all personnel involved in nuclear activities in the host country.

86

c) The regulator has officially documented expectations in the 'National Objectives, Requirements and Model Standards' for the establishment of a security culture programme within the civil nuclear industry in the host country.

d) The cooperative effort made by the competent authority and all other interested parties to promote and enhance a strong security culture is commendable. This contributes significantly to the existence of an effective nuclear security regime.

e) The competent authority has established a community of security professionals within the nuclear sector by organizing security workshops and round tables to develop security culture in the nuclear industry.

II–1.4. Knowledge management

II–1.4.1.   Sharing regulatory experience feedback

The need for an effective system for sharing regulatory experience feedback has been highlighted even by Members States with a mature nuclear power programme. Embarking countries can benefit a lot by these unique examples at a very early stage.

a) "The CNSC had a very comprehensive system for collecting, analysing and sharing regulatory experience feedback. The regulatory experience feedback was disseminated similarly as operating experience feedback. The CNSC shared its experience on the management of regulatory experience feedback actively with other domestic and international organisations.

b) The CNSC used regulatory experience feedback through a lessons-learned approach (regulatory oversight and support) that was well ingrained across its organization. CNSC had a comprehensive system for collecting, analysing, and sharing regulatory experience feedback. The Commission had several policies and practices that fostered the use of regulatory experience for the continuous improvement of the regulatory framework, such as the Strategic Planning Framework and the Regulatory Framework Steering Committee (RFSC). CNSC used many sources for regulatory and operating experience feedback which enabled the Commission to implement the necessary improvements. These sources included inspection reports, audits, evaluations, self-assessments, external peer reviews and conferences, MOUs (Memoranda of Understanding) with fellow international regulators, Regulatory Framework Consultations etc.

c) The CNSC did not only learn from nuclear regulators, but also from non-nuclear ones by performing its own assessment of the incident to draw lessons learned from it. CNSC also worked together with the relevant authorities to assess common deficiencies and, in a few cases, CNSC performed joint inspections with the relevant authority.

d) CNSC participated in the Community of Federal Regulators (CFR) in recognition of the opportunities to learn from other Canadian regulators as well as to share information and lessons learned.

e) CNSC had a MOU with US NRC to share regulatory experience e.g. on package design approval." [II-10]

f) A working group has been established, comprising experts from various governmental organizations on a federal level, for the comprehensive sharing of relevant information about their activities and topical knowledge in the field of nuclear security, so that all involved institutions maintain updated knowledge and can better coordinate their work.

g) The regulatory body efficiently and pro-actively participates in global sharing of information on nuclear security incidents. Such attitude highly improves common understanding of existing threats globally.

II–1.5.   Emergency preparedness and response

Use of performance indicators for emergency preparedness, assigning the responsibilities and necessary training for involved stockholder are good examples to be followed by Member States to meet the EPR challenges.[3]

a) The NRC introduced performance indicators for emergency preparedness, e.g. Drill/Exercise Performance; Emergency Response Organization Drill Participation; Alert and Notification System Reliability, which are evaluated every three months.

b) The procedures developed and implemented by NRC establish clear responsibility of assignments and communication channels to allow an effective management of its internal interfaces between safety, security, and emergency preparedness.

c) Vietnam Agency for Radiation and Nuclear Safety (VARANS) recognized that good knowledge of counterparts is essential for the effective and efficient work in the group. Therefore, Vietnam Agency for Radiation and Nuclear Safety provided basic training on radiation protection and emergency preparedness for the provincial officials, who will take part in the working group, which is going to draft provincial radiological emergency plan.

II–1.6.   Interfaces, communications and involvement of interested parties and licensee(s)

Use of modern IT tools for involvement of interested parties and licensee, public involvement during licence renewal process, enhancing the credibility of regulatory functions and processes through public communication and development of effective communication, consultation, and coordination procedures for complex security issues are some of the good practices selected form IAEA peer review missions to develop an effective communication mechanism.

a) The State Office for Nuclear Safety (SÚJB) was actively engaged in communication with the public through a Web Conference, on which answers were supplied to any relevant question from the public in a timely manner and in a language adapted to laymen.

b) The NRC licensing process, and in particular the licence renewal process is carried out in a very transparent manner, providing opportunities for hearing and public involvement. Several meetings are held in the vicinity of the plants to provide the public with information on the licence renewal process, solicit input on the environmental review, and to provide the results of the NRC's inspections.

c) The country's national response to the Fukushima accident was well-coordinated and addressed key areas in a short timeframe. National environment radiation monitoring was reinforced, contamination of goods and people was monitored at airports and

---

[3] The good practices compiled in Sections II–1.5, II–1.6 (a), II–1.6 (b), II–1.6 (c) and II–1.7 were taken from the IRRS database, which is maintained by the IAEA Division of Nuclear Installation Safety, Regulatory Activities Section.

harbours, public concerns were addressed by significant communication involvement, and cooperation with Japan was conducted through staff support and technical meetings. The swift launch of the Special Safety Inspection process led to the prompt identification of first measures to improve safety. As part of the response to the implications of the Fukushima accident, the exceptional involvement of external experts in the Special Safety Inspection further enhanced the transparency and further reinforced the credibility of the inspection process, while promoting information sharing with interested parties.

d) The regulatory body has been very proactive in working with multiple national organizations that are competent authorities in areas interrelated with physical security of nuclear facilities, nuclear materials, and radioactive sources. This has resulted in excellent collaboration and cooperation, resulting in considerable progress being made on some very sensitive and complex security related issues.

e) The International Physical Protection Advisory Service (IPPAS) mission's team noted that in addition to expected information defining the roles and responsibilities of the respective parties, the MOUs addressed the parties' specific roles and responsibilities with respect to the physical protection regime. The MOUs establish procedures and processes to ensure effective communication, consultation, and coordination between the parties in carrying out their respective roles and responsibilities.

II–1.7.   International cooperation

The regulatory body of an embarking country may benefit a lot from international cooperation during the early phase of its development.

a) Vietnam Agency for Radiation and Nuclear Safety is very engaged in the framework of international cooperation to gain as much experience as possible. It cooperates with, and has concluded bilateral agreements with, some countries that have developed nuclear power programmes worldwide and in the region. These activities support and complement the statutory requirement to incorporate international best practices and experience into regulatory decisions.

b) The NRC's information exchange programmes, and its active participation in the multilateral and bilateral cooperation programmes are providing a strong contribution to worldwide development of nuclear safety practices and to dissemination of knowledge to other countries.

REFERENCES TO ANNEX II

[II-1]   INTERNATIONAL ATOMIC ENERGY AGENCY, IRRS Good Practices, https://www.iaea.org/services/review-missions/integrated-regulatory-review-service-irrs/irrs-good-practices.

[II-2]   INTERNATIONAL ATOMIC ENERGY AGENCY, IRRS Good Practices, https://www.iaea.org/sites/default/files/21/01/organizational-structure-of-the-regulatory-body-and-allocation-of-resources.pdf.

[II-3]   INTERNATIONAL ATOMIC ENERGY AGENCY, IRRS Good Practices, https://www.iaea.org/sites/default/files/21/01/staffing-and-competence-of-the-regulatory-body.pdf.

[II-4]   INTERNATIONAL ATOMIC ENERGY AGENCY, IRRS Good Practices, https://www.iaea.org/sites/default/files/documents/review-missions/irrs_sweden_follow-up_report.pdf.

[II-5]   INTERNATIONAL ATOMIC ENERGY AGENCY, IRRS Good Practices, https://www.iaea.org/sites/default/files/21/01/regulations-and-guides-for-nuclear-power-plants.pdf.

[II-6]   INTERNATIONAL ATOMIC ENERGY AGENCY, IRRS Good Practices, https://www.iaea.org/sites/default/files/21/02/generic-issues-module-6.pdf.

[II-7]   INTERNATIONAL ATOMIC ENERGY AGENCY, IRRS Good Practices, https://www.iaea.org/sites/default/files/21/01/inspection-of-nuclear-power-plants.pdf.

[II-8]   INTERNATIONAL ATOMIC ENERGY AGENCY, IRRS Good Practices, https://www.iaea.org/sites/default/files/21/01/management-system.pdf.

[II-9]   INTERNATIONAL ATOMIC ENERGY AGENCY, IRRS Good Practices, https://www.iaea.org/sites/default/files/21/01/culture-for-safety.pdf.

[II-10] INTERNATIONAL ATOMIC ENERGY AGENCY, Integrated Regulatory Review Service (IRRS), https://www.iaea.org/sites/default/files/21/01/sharing-of-operating-experience-and-regulatory-experience.pdf

ANNEX III.
LESSONS LEARNED FROM EXPERIENCE WITH THE COVID-19 PANDEMIC AS A REFLECTION ON COPING WITH UNFORESEEN SITUATIONS

III–1.1. Challenges experienced in the conduct of regulatory activities during pandemic situation

The COVID-19 pandemic, in addition to affecting the health and lives of millions of people, disrupted many aspects of everyday life due to the widespread adoption of restrictions on travel and on face-to-face interactions between people.

These disruptions resulted in nuclear regulatory bodies, in common with other organizations, facing multiple operational challenges including curtailment of travel and in-person working, absences of staff due to illness or caregiving to children and relatives, and the need to adopt remote working methods.

Not all staff were able to work from home due to a lack of IT infrastructure (lack of personal computers, poor or overloaded internet connections in some areas), and because some activities were not suited to be transferred to a home environment due to confidentiality of documents. For those who were able to work remotely from home, there were additional challenges due to their own private and family situation.

For instance, when conducting meetings under the pandemic situation, there were additional technical and administrative challenges:

- Difficulties in accessing workplace, offices, venues of meetings and seminars, etc.;
- Difficulties in obtaining information from management and non-management resources. Whereas, during face-to-face meetings, the vendors frequently share, show and produce the documents to support their arguments, they may be less willing to share electronic copies of proprietary documents, hampering the meeting;
- Ineffective communication through video conferencing. While it is easy to communicate viewpoints during face-to-face meetings using verbal and non-verbal means, the use of video links tends to impair interpersonal interaction among participants and results in less effective meetings which can take longer to reach resolution;
- Limitation of data transfer and internet connection disruption.

Meetings with the licensees were held remotely, too, and inspections and site visits were reduced to the necessary minimum. Due to these measures, communication both within the regulatory body and between the regulatory body and the licensee decreased in some cases. All 'unofficial' communication that previously took place (e.g. on the margins of meetings at coffee breaks) was almost completely stopped. This posed a challenge to the knowledge transfer both within and between the organizations.

III–1.1.1. Suggested approaches

Although the COVID-19 pandemic presented a unique situation associated with a particular disease, the recurrence of similar disruption of societal functioning needs to be anticipated and prepared for. A new crisis may emerge from different circumstances, such as the spread of other forms of disease or other events such as a widespread electrical power outage, and severe weather. Preparatory actions by the regulatory body, the licensees and other organizations having responsibilities for nuclear safety, include the following:

- Experience and good practices for ensuring knowledge transfer while working remotely are being created and are becoming available worldwide and not only within nuclear community. The regulatory body may follow the development and adopt appropriate good practices.

- The regulatory body may consider establishing some recommendations to its staff for working in the regulatory body's premises at certain times, even if remote working is allowed also in the future.

- Reliable and secure virtual means of communication for discussions and sharing knowledge need to be established.

- All the necessary documentation required to justify or support nuclear safety ought to be available for remote virtual working venues.

- Use of innovating technologies in performing some of the core regulatory functions for example during inspection and review and assessment.

- The regulatory body has to verify that measures are in place to cope with widespread emergency situations.

ABBREVIATIONS

ARPANSA	Australian Radiation Protection and Nuclear Safety Agency
BAPETEN	Badan Pengawas Tenaga Nuklir (Indonesia)
CNS	Centre for Nuclear Safety (Pakistan)
CNSC	Canadian Nuclear Safety Commission
DBT	design basis threat
DNSRP	Directorate of Nuclear Safety and Radiation Protection (Pakistan)
EPR	emergency preparedness and response
FSAR	final safety analysis report
HAEA	Hungarian Atomic Energy Authority
HOF	human and organizational factors
MCR	main control room
MOU	memorandum of understanding
NISAS	National Institute of Safety and Security (Pakistan)
NNR	National Nuclear Regulator (South Africa)
NNSA	National Nuclear Safety Administration (China)
NPP	nuclear power plant
NRC	Nuclear Regulatory Commission (United States of America)
NSC	Nuclear Safety Centre (China)
NSTC	Nuclear Security Training Centre (Pakistan)
OEF	operating experience feedback
PAEC	Pakistan Atomic Energy Commission
PNRA	Pakistan Nuclear Regulatory Authority
PSAR	preliminary safety analysis report
PWR	pressurized water reactor
SAR	safety analysis report
SSCs	structures, systems and components
STUK	Radiation and Nuclear Safety Authority (Finland)
TSO	technical support organization

CONTRIBUTORS TO DRAFTING AND REVIEW

Ayub, M.	Pakistan Nuclear Regulatory Authority, Pakistan
Grant, I.	Consultant, Canada
Horvath, K.	International Atomic Energy Agency
Hussain, T.	International Atomic Energy Agency
Khaliq, M.	International Atomic Energy Agency
Kubanova, I.	International Atomic Energy Agency
Horvath, K.	International Atomic Energy Agency
Proudfoot, B.	Office for Nuclear Regulation, United Kingdom
Shah, Z.A.	International Atomic Energy Agency
Shahzad, M.	International Atomic Energy Agency
Stefanka, Z.	Hungarian Atomic Energy Authority, Hungary
Tuomainen, M.	Radiation and Nuclear Safety Authority, Finland

Consultants Meetings

Vienna, Austria: 26–29 March 2018; 17–21 June 2019; 10–14 February 2020;
9–13 May 2022

IAEA

International Atomic Energy Agency

No. 26

ORDERING LOCALLY

IAEA priced publications may be purchased from the sources listed below or from major local booksellers.

Orders for unpriced publications should be made directly to the IAEA. The contact details are given at the end of this list.

NORTH AMERICA

Bernan / Rowman & Littlefield

15250 NBN Way, Blue Ridge Summit, PA 17214, USA

Telephone: +1 800 462 6420 • Fax: +1 800 338 4550

Email: orders@rowman.com • Web site: www.rowman.com/bernan

REST OF WORLD

Please contact your preferred local supplier, or our lead distributor:

Eurospan Group

Gray's Inn House

127 Clerkenwell Road

London EC1R 5DB

United Kingdom

Trade orders and enquiries:

Telephone: +44 (0)176 760 4972 • Fax: +44 (0)176 760 1640

Email: eurospan@turpin-distribution.com

Individual orders:

www.eurospanbookstore.com/iaea

For further information:

Telephone: +44 (0)207 240 0856 • Fax: +44 (0)207 379 0609

Email: info@eurospangroup.com • Web site: www.eurospangroup.com

Orders for both priced and unpriced publications may be addressed directly to:

Marketing and Sales Unit

International Atomic Energy Agency

Vienna International Centre, PO Box 100, 1400 Vienna, Austria

Telephone: +43 1 2600 22529 or 22530 • Fax: +43 1 26007 22529

Email: sales.publications@iaea.org • Web site: www.iaea.org/publications

23-05463E